Ed Yong is an award-winning science v
biological can be found on his blog Not Exactly
regularly among the pages of New Scientist. In 2002, he left Cambridge University with a Natural Sciences degree in hand and joined Cancer Research UK, initially as a research student and subsequently as a Health Information Manager. In 2007, he won the Daily Telegraph's Science Writer Award and has since freelanced for New Scientist, Nature, SEED, the Economist and the Daily Telegraph. Ed lives in London with his wife. He finds writing about himself in the third person strange and unsettling.

NOT EXACTLY ROCKET SCIENCE

A year of amazing discoveries

Volume 1
2008

Ed Yong

Not Exactly Rocket Science

First published by Lulu in 2008

ISBN 978-1-4092-4228-4

For Alice

Contents

Contents

Introduction

Introduction

Amazing scientific discoveries are being made all the time but to the vast majority of people, they go unnoticed. Science has a reputation for being incredibly complicated, an endeavour that is beyond the grasp of the everyman. It is a false charge. Science can indeed be complicated but it should never have to be impenetrably so.

It is fair to say that scientific knowledge tends to be shrouded in jargon and elitism, with papers being hard to both access and understand. That is a shame, for the knowledge they lock away is both fascinating and enlightening, a public resource that should be freely available to all. The mainstream media takes a stab at translating it but, all too often, simply replaces the technical jargon with a glaze of sensationalism and confusion. When you read a science story in a newspaper, you are usually reading the result of a lengthy production line. It starts, fairly obviously, with the scientists themselves publishing their work in a journal or presenting it at a conference. A press officer then crafts a press release from the paper, explaining salient points in a non-technical way for the benefit of journalists, who typically lack any scientific background to speak of and write their story using the press release. Ultimate control of the words then passes to one or more editors before going to a subeditor, who will trim it and slap on a headline of their own concoction.

With so many steps, it should come as no surprise that the potential for errors creeps in. The analogies are plentiful: a game of Chinese whispers where the message is increasingly distorted; the repeated use of a tape recorder where making copies of copies of copies leads to a loss of quality; indeed, the very replication of genetic information, where every round introduces the possibility for mutations and mistakes. Sometimes, excellent writing escapes these problems, but all too often, you will find technical errors, exaggerated results, and false promises of practical application just around the corner. Context is stripped away and the language is one of certainty. Words like 'proof' and 'official' tend to creep in. It gives the impression of people in an ivory tower laying out facts that must be believed as absolute truth.

But read a good scientific paper and a very different scene presents itself. You will find a landscape of cautious conclusions, acknowledged controversies and controlled expectations. The implications of the research run secondary to the results themselves and, most importantly, the route that led to the results. Authors will pay respect to the huge amounts of work that led up to the latest findings, the giant shoulders that new discoveries are built on. You will read about the careful controls that researchers build into their experiments and the measures they take to account for alternative explanations to their results. Limitations and controversies are freely admitted and every answer throws up scores of fresh questions.

It is these aspects that are sorely needed to make science come alive, to turn it from a stuffy and alien endeavour into one that is fascinating, humble and all-too-human. This is the sense that I wish to convey to people who do not have a scientific background themselves, and my reason for writing Not Exactly Rocket Science. All the pieces on the website, and therefore in this book, have been written from primary papers; they are an undiluted translation of results. And all have been published between October 2007 and October 2008.

Chapter One deals with animals and the amazing adaptations that they use to survive. In Chapter Two, we look at the process of evolution and how it shapes everything from genes to species to languages. Chapter Three deals with ecology and the environment – the ways in which living things interact and affect one another. We travel down the microscope in Chapter Four to the hidden world of bacteria and viruses and the adaptations used by the tinier side of life. Over the next three chapters, we turn our attentions back to ourselves, starting with a look at our bodies, how they evolved, what happens when they fail and how they can be fixed. In Chapter Six, we consider our behaviour and how the field of psychology is constantly challenging our assumptions about the ways in which we think and behave. Chapter Seven deals with the world of brains and neurons and will shed even more light on how our minds work. And finally, Chapter Eight looks back into the past, at species that are no longer with us.

Some of the topics covered will certainly be of great importance to humanity over the coming years – stem cells, climate change and falling biodiversity, to take some examples. Understanding these issues will become increasingly vital as they start to take centre stage, and a growing number of ethical and political decisions rely on appreciating them. But the vast majority of subjects in this book are bereft of such practical value, and deliberately so. Their benefit lies simply in the joy of extra knowing – in expanding our view of the world around us, telling us about our origins and perhaps most importantly, laying out what we still have left to discover. To take just one example, last year saw the discovery of a virus called Sputnik that infects other viruses. It was a chance find and it is likely that there are many more so-called 'virophages' out there. This is an entire class of life that until now was completely unknown to us and it is indicative of the fact that we are just scratching the surface. This, I feel, is part of the true value of scientific discoveries – in an age where it seems like all the answers are a quick Google search away, they cultivate a sense of inquiry, remind us of the depths of our ignorance and provide us with routes for exploring them. That is what this book seeks to celebrate.

Ark of the amazing

Magnetic cows, binge-drinking shrews, defensive cloning, scared elephants and roach-walking wasps

Chimps outsmart students

We humans are not used to having our intelligence challenged. Among the animal kingdom, we hold no records for speed, strength or size but our vaunted mental abilities are unparalleled. That is, until now. New research from Kyoto University shows that some chimps have a photographic memory that puts humans to shame. Sana Inoue and Tetsuro Matsuzawa have found that young chimps have an ability to memorise details of complex images that is literally super-human. One particular boffin chimp, Ayumu, outperformed university students in memory tasks where they had to rapidly memorise numbers scattered on a touch screen and press them in numerical order.[1]

This is the first time that an animal has outmatched humans in a mental skill. Over the past few years, animals have displayed a variety of abilities that were once considered to be uniquely human. Jays can plan for the future, rats can reflect on their own knowledge, chimps have their own cultures and crows can combine tools together. But in all these cases, the animals merely showed that they could perform similar types of mental feats to us; they never challenged our abilities in terms of complexity or scale. Simply put, a crow may be able to combine tools together, but it is never going to be able to engineer a computer.

Three years ago, Inoue and Matsuzawa started to teach three pairs of chimp mothers and children to play with numbers. One of the mums, Ai, was the first chimp to learn to use Arabic numerals to accurately number sets of real-life objects, but the other five had never done memory tasks involving numbers. Using a touch-screen computer, the researchers eventually trained all the chimps to touch combinations of numbers from one to nine in numerical order. When the youngsters reached their fifth birthday, Inoue and Matsuzawa taught them a more complicated task. When they touched the first digit, the others were replaced with white squares, and they had to rely on their memory to press the squares in the right order. The young chimps took to this task particularly well and amazingly, they finished the task more quickly than human adults.

Ai's son, Ayumu, emerged as the class star and went on to the next challenge. He had to touch a white circle to bring up a random selection of five numbers, which were quickly replaced with white squares. Again, he had to rely on memory to press the hidden numerals in the right order. When the numbers were flashed for two-thirds of a second, Ayumu's skills were the equal of six university students who pressed the right sequence 80% of the time. If the numbers were displayed for just a fifth of a second, the students could not cope. They did not have enough time to make a single saccade, the small eye flickers that we make when we scan a page or image. Without the luxury of exploring the screen, the students only answered accurately 40% of the time. Ayumu, on the other hand, was not fazed and maintained his earlier high scores.

Inoue and Matsuzawa claim that Ayumu has a photographic or "eidetic" memory that allows him to keep an accurate, detailed image of a complex scene in a very short time. His ability suggests that we should not underestimate the mental abilities of other animals, and our chimp cousins in particular. However, chimps have not quite won the memory battle just yet. A large proportion of human children have photographic memories but the ability fades away as they age. Ai's performance, which was worse than that of her son and the human students, suggests that chimps might experience the same decline. If that is the case, pitting a chimp in the prime of his memory like Ayumu against human teens might be an unfair contest. It would be very interesting to see if he could still out-remember human children whose eidetic memories are still in their prime.

Cattle compasses

For centuries, farmers have known that their livestock not only gather in large herds but also tend to face the same way when grazing. Experience and folk wisdom offer several possible reasons for this mutual alignment. Perhaps they stand perpendicularly to the sun's rays in the cool morning to absorb heat through their large flanks, or maybe they stand in the direction of strong winds to avoid being unduly buffeted and chilled. But cows and sheep do not just line up during chilly spells or high wind. Their motivations for doing so during warm, pleasant and unremarkable weather, or indeed in the dead of night, remained a mystery, that is until Sabine Begali from the University of Duisburg-Essen, Germany decided to spy on aligned herds of cows and deer using satellite images from Google Earth. The images revealed a striking behaviour that had been going unnoticed for millennia, right under the noses of herdsmen and hunters – their herds were lining up in a north-south line like a living compass needle. Influenced by a magnetic sense that has only just become apparent, their default point of reference is not the source of wind or the angle of the sun, but the Earth's magnetic poles.[2]

With Google Earth's satellite images at their disposal, Begali's team spied on a massive sample of cattle across six continents, from South Africa to India to the UK. They recorded the positions of over 8,500 individuals at more than 300 sites, including a range of different breeds, altitudes and times. Their global cattle census showed that the animals oriented themselves along a north-south axis so consistently that the odds of them doing so by chance were less than one in a hundred thousand. The animals do not quite point towards the North Pole, but instead face slightly off it in the direction of magnetic north. As this position (known as magnetic declination) changes across the face of the planet, so too does the direction that local herds prefer to face.

Deer aligned themselves to the magnetic poles even more strongly than cattle. Begali's team travelled to more than 200 locations in the Czech Republic and observed the positions of almost 3,000 roe deer and red deer in the field. When the animals had moved, the team also recorded the alignment of the body prints that the resting deer had left behind in snow. Again, their bodies faced magnetic north. Even their heads tended to gaze in that direction, although less predictably so – after all, they do need to look around for predators like lynxes.

Begali says that the study's large sample size has helped the team to rule out other explanations for this uncanny alignment. Certainly, the typical answers of sun and wind seem less applicable in the face of such standard behaviour across the globe. Unfortunately, Google Earth images are not time-stamped with enough accuracy to be able to compare the shots with weather data. But the fact that the photos exist at all suggest that they were taken on cloudless days, and the shapes and positions of shadows suggest that the sun was not too bright. If cattle were lining up primarily in response to gusty winds, the majority of the 308 locations that the team sampled must have been experiencing high winds at the time and specifically in a northerly-southerly direction. That is very unlikely, given that westerly breezes dominate the Northern hemisphere and south-eastern trade winds rule the Southern. If that weren't enough, wind atlases show that the prevailing winds in the countries in question vary throughout the year and if there is any directional consensus, it is a westerly one. So much for the wind; Begali's data rule out a major influence from the sun too. The satellite images also recorded the position of the animals' shadows and these revealed that by and large, they weren't getting their bearings from the angle of the sun. And the fact that deer still faced the same way at night also argues against the sun's involvement. There is a third alternative – some animals including insects and migratory birds can sense polarised light from the sun and use it to navigate. However, there is no evidence that cows or deer have the right retinas for picking up polarised light and given that they are partially active at night, they do not fit the typical profile of polarised light-users, who are usually only active by day.

With these alternatives effectively discarded, only one explanation remains, and it is the simplest one – the cows and deer were using the Earth's magnetic field as their guide. They are not alone; many other animals, including flies, bees and goldfish, naturally line up like little compasses when there is not anything around to disturb them. So why do it? That is still a mystery. One slightly leftfield possibility is that keeping magnetic fields in symmetry about the axis of your body could affect certain bodily processes. There are tantalising bits of evidence to back this up – in humans, the time it takes to drift into deep REM sleep differs depending on whether we're facing a north-south direction or an east-west one. So does the electrical activity in our brains. Alternatively, it could just be that orienting yourself in a constant direction makes it easier to get your bearings when travelling long distances or making quick getaways. Indeed, both cattle and deer are naturally social animals that travel large distances across landscapes that are often bereft of landmarks. On such featureless terrain, a magnetic sense could be good for navigation, as turtles and birds can testify.

It is not clear whether the cows are actively aware of their ability. It is certainly possible for creatures to have the capacity to detect magnetic fields, known as magnetoreception, without consciously sensing them, which is known as magnetoperception. Magnetic senses are one of the most enigmatic of animal abilities. Among our fellow mammals, only a few rodents and one species of bat are known to use internal compasses. Some believe that horses, dolphins and whales use the same trick but that has been very hard to prove. Large groups of these animals do not lend themselves to careful laboratory experiments. Begali's innovation was in using satellites to turn the entire planet into a natural laboratory. Now, it is time for others to follow up on her results.

Roach-walkers, head-bangers and body-snatchers

The vast majority of wasps are "parasitoids", animals that practice the grisly art of body-snatching. They lay their eggs in the bodies of other living animals to provide their newly hatched grubs with a fresh supply of meat. Like HR Giger's alien, the full-grown larvae then burst through their host's skin, usually killing it in the process. The jewel wasp (*Ampulex compressa*) is one of these body-snatchers, and its stunning colours belie its gruesome habits. Its grubs feed on the bodies of cockroaches supplied by their mother. When a female wasp finds a roach, she stings it twice – once in its mid-section to immobilise its front legs, and the second time directly into its brain. There, she pumps in venom that stupefies the roach and changes its behaviour. It is not paralysed but it moves sluggishly and shows no desire to flee from danger. In this befuddled state, the jewel wasp can grab the roach by its antennae and walk it around like a dog on a leash. The wasp leads its roach to its nest, where it seals it up and lays an egg on its belly. Even as the larva hatches and starts to eat the roach alive, the hapless insect does not struggle or fight.

Now, Ram Gal and Frederic Libersat from Ben-Gurion University in Israel have discovered how the wasp's venom keeps its victim so sedate but otherwise mobile and healthy – it is an incredibly precise tool that specifically reduces the cockroach's motivation to walk.[3] Gal and Libersat placed stung cockroaches into a chamber where half the floor could be electrified. For a normal cockroach, a shock that is strong enough to make its leg muscles twitch is usually strong enough to make it walk to safety. When the roaches are first stung, they behave in much the same way, but their responses soon change. An hour later, it takes almost five times the previous voltage to make them scamper even though their leg muscles contract in a normal way. After four hours, the shocks need to be 10 times

stronger. It also took as many as four shocks in a row to reliably get them moving. In the experiment, their responses only returned to normal after 72 hours when the venom's effects started to wear off. Of course, in the wild, it would be far too late for them by that point.

These results show that it takes a lot more to get the stung roaches to move on their own accord. As Gal and Libersat put it, the stung cockroaches have "a deficit in 'reaching the decision' to walk". They are also less likely to keep on moving once they have started. Gal and Libersat demonstrated this by placing the roaches in a water-filled cylinder, a potentially fatal predicament that they need to escape within minutes. Normally, cockroaches start swimming as soon as they hit the water, and spend about 90% of their time trying to escape from drowning. And while the vast majority of the stung cockroaches also started to swim, they did not keep at it for long, kicking for just 10 seconds in every minute. It was not that they were simply tired; when Gal and Libersat removed the slackers from the cylinder and turned them upside-down, they tried to right themselves by kicking out with their legs, with just as much vigour as normal cockroaches.

The fact that stung cockroaches could still flip themselves back on their feet shows that the jewel wasp's venom does not affect their general motor skills. When they try to right themselves, their muscles show the same degree of activity as those of normal cockroaches in the same dilemma. Nor did the venom affect the roaches' ability to fly. When Gal and Libersat blew air over the wind-sensitive hairs on the insects' back ends, their wing muscles contracted at normal strength and frequency. Gal and Libersat suggest that the wasp's venom could affect certain signalling chemicals such as octopamine and dopamine that affect the movements and motivations of insects. Indeed, the group has previously shown that the zombie cockroaches can be restored to their active ways by injecting them with octopamine. Regardless of the method, it is clear that the wasp's venom is a precision weapon that does not just indiscriminately target the roach's responsiveness or its ability to move. Instead, it is so well matched to the brain of its prey that it only affects the specific neural circuits that are involved in walking.

The jewel wasp turns cockroaches into zombies-on-a-leash, but other wasps use a different strategy. One, for instance, recruits the caterpillars of geometer moths as bouncers. Human bodyguards do risky work and they are usually well-paid for their troubles but the moth caterpillars get no such reward. Against their will, they are recruited to defend the developing young of a parasitic wasp, and their only 'reward' is to be eaten inside out by the larvae of their attacker. For one species, *Thyrinteina leucocerae*, the ignominy does not end there. It is targeted by a *Glyptapanteles* wasp that, on a single pass, can lay as many as 80 eggs onto a hapless caterpillar. Two weeks later, the larvae burst through their host's skin but despite its injuries, the caterpillar remains alive and stays near the hatched grubs as they spin their pupae and turn into adults. It never moves and it never feeds. All it does it violently swing its head in response to nearby movement. After the adult wasps fly off, it eventually dies.

Amir Grosman from the University of Amsterdam has found that the caterpillar's strange behaviour is all part of the manipulations of the wasp.[4] Its last act is to defend the very grubs that spent the last two weeks killing it, playing the role of bodyguard as well as incubator. Together with Dutch and Brazilian colleagues, Grosman showed that caterpillars that were incubating wasp grubs were just as active as unaffected ones. However, after the grubs hatched, the majority of the host caterpillars froze completely, stopped feeding and reared up onto their hind legs. These zombie-like sentinels only moved in the presence of stinkbugs, which will eat both caterpillars and wasp pupae. When Grosman placed stinkbugs near the caterpillars, almost all of the parasitized ones lashed out with violent swings of their heads. These head butts are not a normal response, for only one of the 20 unaffected caterpillars reacted to stinkbugs in the same way. They are, however, an effective defence. In about 60% of encounters with the head-banging caterpillars, the bugs either gave up or were knocked off their twigs. In natural conditions, the caterpillar's vigil ensures that more wasp pupae survive. When Grosman removed these unwitting guardians, the death rate among the pupae doubled. Some were eaten and a few, in an ironic twist, became hosts themselves to other wasp grubs – a case of so-called "hyperparasitism". Clearly, the young wasps benefit from their warden's actions but the caterpillars themselves get nothing. Every last one of them was dead within a week after the pupae opened and the adult wasps emerged.

The ability to change the behaviour of its host is a common strategy in a parasite's playbook, and this is just one case among many. Another wasp, *Hymenoepimecis argyraphaga*, induces its spider host to spin a radically different type of web to support the wasp's cocoon. The deranged spider is eventually sucked dry by the grub hanging onto its body. Snails infected with the *Leucochloridium* fluke neglect their shade-seeking instincts and crawl onto brightly lit leaves. There, the fluke extends pulsating, brightly coloured sacs of eggs into the snail's antennae, where they catch the attention of the fluke's next host – birds. And one brain parasite, *Toxoplasma gondii*, makes rodent hosts less fearful of the smell of cats, which are its final host. It may even have affected human culture – the prevalence of *T.gondii* in a country is a reasonably good indicator of how neurotic people from that country tend to be.

Returning to the wasps, how do they make the caterpillars change its behaviour so radically? The weird change only happens about two weeks after the female wasp originally filled the caterpillar with eggs, so she is unlikely to play any role in it. The hatching process is not the trigger, because caterpillars that are artificially damaged do not behave in the same way. And the pupae do not secrete any mind-altering chemicals because placing them next to uninfected caterpillars has no effect. The most likely scenario is altogether more amazing. When Grosman dissected guardian caterpillars, he found that not all the wasp larvae leave their hosts – one or two stay behind and are still active. Grosman thinks that it is these hangers-on that manipulate the half-dead caterpillar. They effectively sacrifice themselves for the survival of their siblings.

Cloning in defence

Many animals have cunning ways of hiding from predators. But the larva of the sand dollar takes that to an extreme – it avoids being spotted by splitting itself into two identical clones.[5] Sand dollars are members of a group of animals called echinoderms, which includes sea urchins and starfish. An adult sand dollar (*Dendraster excentricus*) is a flat, round disc that lives a sedate life on the sea floor. Its larva, also known as a pluteus, is very different – a small, six-armed creature that floats freely among the ocean's plankton. A pluteus cannot swim

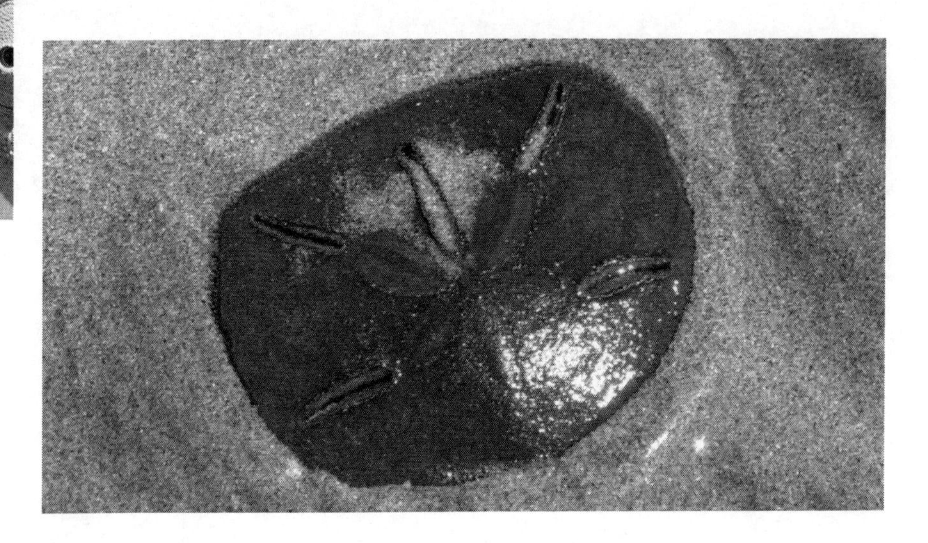

quickly, so there is no escape for one if it is attacked by a hungry fish. Instead, Dawn Vaughan and Richard Strathmann from the University of Washington discovered that the pluteus relies on not being spotted in the first place.

They exposed 4-day-old larvae to water infused with mucus from the skin of a potential predator – the Dover sole. Within 24 hours, every single larva that was exposed to the mucus has grown a small bud that eventually detached and developed into a second larva, genetically identical to its parent and smaller in size. In contrast, larvae that were exposed to untouched seawater did not divide. The tactic presumably works because hunters find smaller larvae harder to spot. As a result of the cloning process, the original larva halves in size. The newborn clones were smaller still and some were no bigger than a sand dollar egg cell, just an eighth of a millimetre across.

Almost all groups of echinoderms clone themselves in response to favourable temperatures and abundant food. In fact, asexual reproduction is a remarkably common tactic among plants and animals and is even used by some very unexpected species. Recently, both Komodo dragons and hammerhead sharks have cloned themselves in captivity (with the dragon actually having its virgin birth in time for the Christmas headlines). Cloning yourself brings several advantages; an animal can rapidly produce large swathes of offspring to make the most of plentiful times and in the long run, asexual reproduction can even allow a species to divert different copies of its genes to new purposes (see page 46) But this is the first example of an animal using a virgin birth to defend itself.

Obviously, the tactic is not meant to protect the young sand dollar against an attack – even the slowest fish would have made a mouthful of the dividing larva within 24 hours. It is more of a long-term strategy, allowing the larva to prepare itself for the presence of predators, as indicated by the smell of mucus in the water. That is not something that adult sand dollars can do. From their position on the sea floor, they have scant information about the risks that their young plutei may face and cannot alter the size of their eggs appropriately. Instead, Vaughan and Strathmann suggest that the cloning tactic allows the larvae themselves to alter their size in response to the risks they sense.

Memories from metamorphosis

The transformation from caterpillar to butterfly or moth is one of the most beguiling in the animal world. Both larva and adult are just stages in the life of a single animal, but are nonetheless completely separated in appearance, habitat and behaviour. The imagery associated with such change is inescapably beautiful, and as entrancing to a poet as it is to a biologist. According to popular belief, within the pupa, the caterpillar's body is completely overhauled, broken down into a form of soup and rebuilt into a winged adult. The chemist Richard Buckmister Fuller once said that "there is nothing in a caterpillar that tells you it is going to be a butterfly." Indeed, as the butterfly or moth flies off into a new world, it is tempting to think that there is no connection between its new life and its old existence as an eating machine.

But not so – the larval and adult stages are not as disparate as they might seem. Adult tobacco hookworms – a species of moth – can remember things that they learned as a caterpillar, which means that despite the dramatic nature of metamorphosis, some elements of the young insect's nervous system remain intact through the process.[6] Using some mild electric shocks, Douglas Blackiston from Georgetown University trained hookworm caterpillars (*Manduca sexta*) to avoid the scent of a simple organic chemical – ethyl acetate. The larvae were then placed in the bottom end of a Y-shaped tube, with the scent of ethyl acetate wafting down one arm and fresh air coming down the other. Sure enough, 78% of the trained caterpillars inched down the odour-free arm. As the caterpillar moulted their way through the larval stage, their aversion to ethyl acetate remained. Blackiston allowed them to pupate and emerge as full-grown moths, before

testing them again, about a month after their initial 'electric' education. Bear in mind that a tobacco hornworm lives for about 30 to 50 days, so a month is very close to its entire lifespan. Amazingly, 77% of the adult moths also avoided the ethyl acetate-scented arm of the Y-shaped tube and the vast majority of these were the adult versions of the same larvae that had correctly learned the behaviour originally. Clearly, the larvae had learned to avoid the chemical and that memory carried over into adulthood.

Even so, Blackiston was careful to account for alternative explanations. For a start, he ruled out the possibility that the adults just did not like the smell of ethyl acetate themselves. Ethyl acetate is not naturally foul-smelling and it is actually rather reminiscent of pear drops. When larvae are exposed to it in the absence of electric shocks, neither they nor the adults they become learn to avoid it. Another possible explanation hinges on the fact that adults emerging from the pupa usually experience a similar milieu of smells to their caterpillar selves. This chemical legacy could explain why adults and larvae react similarly to some odours. But when Blackiston applied ethyl acetate gel to the pupae of untrained caterpillars, the adults did not shrink away from the chemical. Nor did washing the pupae of trained to get rid of any lingering traces of ethyl acetate have any effect.

Blackiston was convinced that some aspect of the caterpillar's nervous system was carried over into adulthood. However, he also found that this only happened if caterpillars are trained at the last possible stage before they pupate – the 'fifth instar'. Any earlier, and the memories do not stick. The fruit fly *Drosophila* suggests why this might happen. In its brain, memories of smells are located in mushroom bodies, brain structures that consist of three lobes. The gamma lobe develops very early while the alpha and beta lobes develop just before the pupal stage. Blackiston thinks that long-lasting larval memories are writ into the alpha and beta lobes, whose neural networks are kept around while the rest of the caterpillar breaks down. If the larvae are too young, these areas have not developed yet and any learned information is stored in the gamma lobe only to be lost when its connections are trimmed back in the pupa.

But why bother? After all, the entire advantage of metamorphosis rests on the very different lifestyles and habitats of caterpillars and moths, which allow them to avoid competing with each other. Nonetheless, moths and butterflies must still return to the right sort of plant in order to lay their eggs and Blackiston suggests that their larva-hood memories may help them to do so.

Monkey be good, monkey feel good

There are some who say that helping others is its own reward, and many biologists would agree. The fact that selfless acts give us a warm glow is evident from both personal experience and neurological studies, which find that good deeds trigger activity in parts of the brain involved in feelings of reward. But feeling food by being good is not just the province of humans – monkeys too get a kick out of the simple act of giving to their fellow simians.

At the Yerkes National Primate Research Center, Frans de Waal's team of scientists have been investigating the selfless side of eight brown capuchin monkeys.[7] Each monkey was given a choice between two differently coloured tokens. Both would earn it a rewarding piece of apple but only one token would net a slice for a second monkey sitting in an adjacent transparent compartment. The chooser would benefit equally no matter what choice they made, so if they were completely cold to the needs of their peers, you would expect them to pick both tokens with equal frequency. Otherwise, the "prosocial" token which benefited a second monkey would be the favoured pick. Everything else begin equal, would the monkeys take the welfare of their fellows into account?

They did. De Waal put the capuchins through three blocks of ten trials and even during the first set, they were more likely than not to choose the token that benefited the other monkey. By the end of the experiment, they were selecting the prosocial token about two-thirds of the time. The capuchins hailed from two different social groups and the team's close knowledge of these cliques allowed them to probe the boundaries of the animals' charitable side. Humans show the strongest degree of empathy towards those closest to them – friends and family – and you would expect the capuchins to behave similarly. Indeed they did, and they were most likely to pick the prosocial tokens when the other monkey was a blood relative. They were also willing to help out unrelated members of the same group but the line stopped at unrelated strangers, who only received a tasty treat during half of the trials.

These sorts of experiments are quite difficult to do while ensuring that the monkeys' responses are truly genuine, so the team went out of their way to rule out other explanations. They did a set of trials while wearing a tinted face-shield, to account for the possibility that the monkeys were biased towards one choice over another because of subtle movements of the experimenters' heads or eyes. And to control for favouritism for tokens in one position, they randomly swapped the position of the two tokens, or gave the monkeys a choice between six jumbled ones. Neither change had any effect on the frequency of the monkeys' giving behaviour.

Perhaps the animals were influencing each other? It is possible that monkeys were only helping out their peers because they feared retribution when they were returned to their enclosures. But on average, the dominant monkeys were actually slightly more likely to be selfless than subordinate ones, which is exactly the opposite pattern you would expect if their charity was really motivated by fear of punishment. The partner monkey hardly ever actively threatened or cajoled the chooser either. Even so, it is clear that the partner's presence was important and that the chooser knew what it was doing when it picked the token that would benefit them both. During these choices, the chooser tended to sit close to their partner, face them and exchange social signals. And if the team slid an opaque screen in front of the second individual so the chooser could not see it, they mostly picked the selfish option.

De Waal says that having ruled out fear of punishment, bias, persuasion and other alternatives, the best explanation for the capuchins' actions is that the act of giving is personally rewarding for them. No matter what decision the chooser makes, it walks away with a mouthful of apple so any benefits gained by picking the prosocial token must be intangible ones. To de Waal, these intangible perks probably resemble the warm glow that we humans get when we help each other out. It all boils down to empathy, the ability to understand someone else's state of mind. This simple aptitude automatically gives an individual an automatic stake in the wellbeing of its peers – watch someone else feel good and you will too. Humans are notable for our strong sense of empathy, but de Waal's capuchins have demonstrated a faculty for it too. And there was another side to their altruism that we might recognise – a contingency on fairness. Capuchins prefer grapes to apples and if the prosocial token led to the partner getting a grape while the chooser just received an apple, they were less likely to be selfless. They were happy seeing their compatriots being rewarded, but not with a gift more extravagant than their own.

Altruism – the selfless concern for others – is one of the hottest topics in evolutionary biology, but so much effort is spent on understanding its origins that very few scientists try to examine the causes of altruism in the here and now. This study is one of the few that have tried to look at the immediate motivations behind selfless acts, aside from their long-term evolutionary benefits. We can draw an analogy to other behaviours. Sex, for example, allows an animal to pass on its genes to the next generation, and such evolutionary explanations are known as "ultimate" ones. But individuals are hardly thinking about that when they are getting it on and most humans actively try to avoid thinking about it! Animals mate because combinations of hormones and nervous impulses drive them into an amorous state – these are the "proximate" causes of the behaviour. In the case of cooperation, the ultimate evolutionary reasons for it include the benefits of helping out related individuals who share many of the same genes, and preserving your own life by helping those who are likely to reciprocate in the future. Neither is a good proximate reason – they can tell us how altruism emerged, but not about a capuchin's reasons for offering food to their peers in the present. If we believe that monkeys help each other in anticipation of a reciprocal hand, we also need to assume that they can predict the likelihood that other monkeys would return the favour. That is a complex mental ability and one we shouldn't just assume exists. To de Waal, the idea that monkeys can empathise with each other is more plausible. It is a trait that they share with us. But not all of us.

The X-Frog

In the X-Men comics, the superhero Wolverine is armed with three sharp claws on each arm. They extend through the skin of his hand, and the resulting wounds are closed by up his superhuman ability to heal. Now, in a bizarre case of life imitating art, scientists from Harvard University have discovered that a group of African frogs use similar weapons. The frogs defend themselves with sharp bone claws on their hind feet but to do so, the animals have to drive the claws through their own skin.[8] It is an extreme defence that is completely unique in the animal world.

The clawed frogs belong to a family called Arthroleptidae that were discovered in Central Africa more than a century ago. At first, people wondered if the claws just stuck through the skin as a side effect of the preservation process or if the frogs may have used them to grip or climb. Their true function as defensive weapons only became clear when naturalists first described actually picking up and handling live animals. Doing so is a mistake, and anyone who makes it is punished with a series of deep, bleeding wounds inflicted by the struggling animal as it kicks out violently with its claws. The ability is well known to the people of Cameroon, who only ever hunt the frogs with machetes or spears.

David Blackburn from Harvard University revealed how the claws work by studying 63 specimens housed in museums. Two genera in particular – *Astylosternus* (the night frogs) and *Trichobatrachus* (the hairy frog) – have well developed claws on eight of their hind toes. Claws are, of course, very common in the animal kingdom and cats also have retractable claws on their feet. But those of the frogs are completely unique for several reasons. For a start, they are made of naked bone and lack the sheath of keratin that covers the claws of all other land-living back-boned animals. The way they work is special too. When the claw is 'at rest', it lies completely within the flesh of the toe, held in place by a small, bony nodule, which is in turn

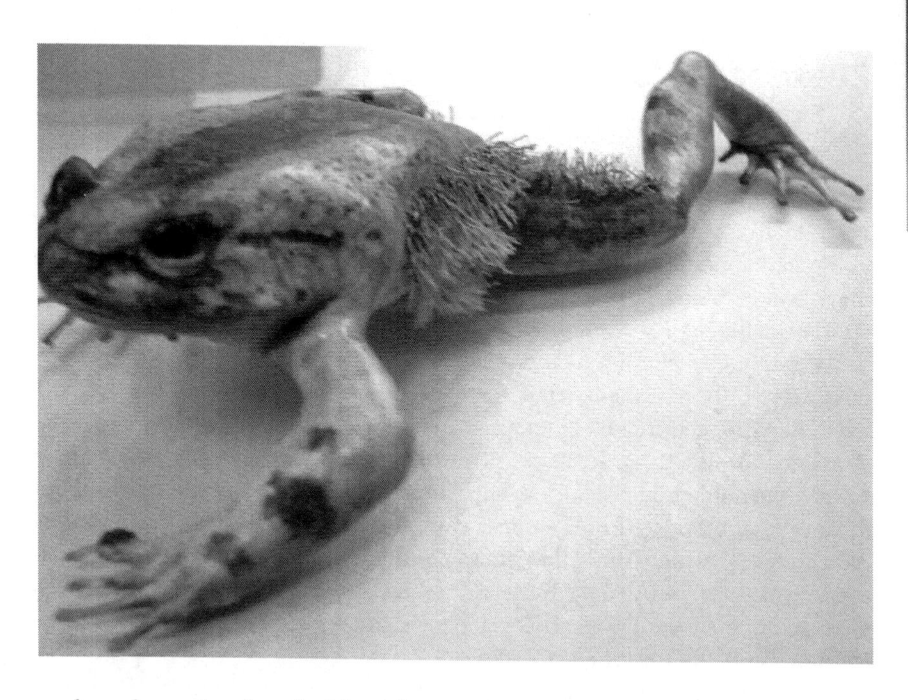

anchored to the frogs' skin. The nodule holds the claw in place, and prevents it from being accidentally extended while the frogs go about their day-to-day business. When the frogs are threatened, they flex a muscle that connects to the lower half of the claw. This severs the connection to the nodule and pulls the claw downward. Its barb-like tip is made of reinforced and thickened bone, which easily pierces through the skin of the toe.

Among back-boned animals, no other claw needs to pierce its way to use. Only salamanders use a comparable technique; some species jab their bony ribs through special sites in their skin that contain poison glands. And unlike the toes of cats, the frogs have no permanent slots through which the claws extend. Every time they are used, they create new wounds. In the X-Men movie, Wolverine, when asked if it hurts to pops his claws, answers, "Every time." One cannot help but think that the same is true for the frogs.

Blackburn is not sure what happens after the frog has deployed its secret weapon. It may be that the animal cannot actively retract the claws and has to wait for them to passively return to the 'cocked' position. Nor do we know if the broken connection between the bony nodule and the claw will regrow. However, many amphibians have extraordinary healing abilities that can even regenerate severed limbs. It may be that the clawed frogs, like their comic-book superhero counterpart, have a 'healing factor' that closes up the wounds that open every time their claws are used.

Bees scare elephants

It is a myth that elephants are afraid of mice, but new research shows that they are not too keen on bees. Even though they fearlessly stand up to lions, the mere buzzing of bees is enough to send a herd of elephants running off.[9] Armed with this knowledge, African farmers may soon be able to use strategically placed hives or recordings to minimise conflicts with elephants.

Iain Douglas-Hamilton and Fritz Vollrath from Kenyan conservation charity Save the Elephants first suspected this elephantine phobia in 2002, when they noticed that elephants were less likely to damage acacia trees that contained beehives. Animals as powerful as the African elephant can go largely untroubled by predators. Their bulk alone protects them from all but the most ambitious of lion prides. But these defences do nothing against African bees, which can sting them in their eyes, behind their ears and inside their trunks. Against these aggressive insects, the elephants are well justified in their caution and local people have reported swarms of bees chasing elephants for long distances.

Lucy King, a graduate student from the University of Oxford confirmed this theory by using camouflaged wireless speakers to play recordings of angry buzzing bees to herds of elephants resting under trees. The buzzing caused almost unanimous alarm. The elephants stopped what they were doing and scanned their surroundings with raised heads, spread ears and swishing trunks. Within 10 seconds of hearing the recording, almost half of the families had fled with their tails in the air, occasionally throwing backwards glances at the speakers. By the 80 second mark, all but one was gone. In contrast, only seven groups scattered when they heard a control recording – a burst of white noise extracted from the recording of a waterfall. The leisurely pace of these groups suggests that they moved out of irritation. In contrast, the elephants that fled from the sound of buzzing moved more quickly, suggesting that they were motivated by fear rather than annoyance. The groups that were buzzed off also moved about 60 metres away from the recording, more than three times the distance that the white noise groups did.

King notes that her study does not show how the elephants come to develop their phobia of bees. Naive individuals may have learnt the lesson the hard way – from being stung – or may learn what to do from watching more experienced adults. The single group that stood its ground suggests that the second theory may be right. This group was unusually small and young for an elephant herd, consisting only of a young 20-year old male, a 14-year old female and her calf. Usually, herds have several older adults and a matriarch who leads them. It could be that none of the three elephants had been stung themselves, and without an experienced leader, they did not know the right response. This just goes to show how important social structures are to elephants, where youngsters learn appropriate behaviours from their elders.

The research team hope that their discovery could be put to practical use. In many parts of Africa, expanding human settlements are pushing elephants into ever-smaller ranges, leading to mounting conflicts between the two species. The pachyderms frequently raid crops causing massive economic losses. Some scientists believe that elephants may even be suffering from a form of post-traumatic stress disorder and be acting out of spite. Fencing the elephants out with electric wiring and fortifications are expensive and difficult to maintain. Bees, on the other hand, could provide a simple and profitable solution and the trio now plan to test this idea using a combination of actual hives and powerful loudspeakers. Strategically placed hives could not only deter marauding elephants, but also produce sellable honey. It is a win-win situation, which is very rare in conservation.

March of the cannibals

A tenth of the planet's population occasionally suffers through devastating famines because small insects fear being bitten in the bum. That is the astonishing message from the latest research on one of mankind's greatest pests – the desert locust. Swarms of locusts can stretch for several hundred square kilometres and each of these harbours up to 80 million hungry sets of mandibles that eat their own body weight in food every day. These plagues are unpredictable but they only form when locust populations reach some sort of critical mass. Desert locusts are two insects for the price of one; at a crowded tipping point, they transform from loners (which are green or brown) into more sociable forms that are red or yellow and prone to swarming. With much of Asia's and Africa's food supplies at stake, researchers are keen to discover what prompts the transformation from disordered groups of solitary locusts to highly organised marches of sociable ones.

Two years ago, an international team led by Stephen Simpson showed that this switch is very rapid. Once groups reach a certain density, individuals that were previously doing their own thing started stepping in line with their neighbours. This sudden coordination is an important step in the genesis of a swarm, but the researchers had still to uncover why the locusts aligned so neatly. Now, working with the same group, Sepideh Bazazi at the University of Oxford has found part of the answer – they march to avoid getting cannibalised by other locusts behind them.[10] All the individuals in a dense group are after the same things – protein, salt and the like. If one stops moving, it risks acting as a source of these nutrients for others behind it. For locust groups, life is about moving with the crowd, or being eaten by it.

Bazazi studied the movements of locusts by making them march around a round arena with a raised dome in the middle, looking rather like a metallic sombrero. Automatic software tracked their speed and positions as they circled the ring. Bazazi found that locusts are far more likely to start moving when they detect other locusts approaching from behind. When she severed the main nerve cord in the insects' abdomens and deprived them of feeling, she found that they were less likely to move than individuals with full feelings in their backsides, and those that did move walked more slowly. This reticence only manifested in groups; solitary locusts that were operated on were no different to their untouched peers, showing that the surgery was not affecting their general movement. It was their ability to sense movement behind them that had been specifically removed, and this inability cost the numbed locusts dearly – they became six times more likely to be injured by cannibalistic attacks.

As a final test, Bazazi used dabs of black paint to block out certain parts of the locusts' fields of vision to show that locusts are more likely to move based on what they see behind themselves than what they see in front of them. Both totally blind locusts and those that could not see behind themselves were half as likely to move as individuals with a full range of vision. Locusts that could see behind but not in front of themselves fell somewhere in between. Putting all these pieces of the puzzle together, Bazazi suggests that locusts in a group start to march forwards to avoid getting bitten by those behind them. The fleeing locusts run into those in front of it, who themselves become a potential meal and have their own impetus to move forward. The result is a vicious cycle, where the movements of just a small number of individuals can set an entire group off into a march. Cannibalism is not the only motivating factor for the locusts – certainly, a need to find more food drives them too – but it could play an important role in pushing the insects to march in the first place. Once this happens, constant contact on their legs turns a march into a swarm. Simpson's lab had previously found that repeatedly touching the hairs on the insects' hind legs causes them to shift from the solitary form to the swarming one within a matter of hours.

Locusts are not the only member of the grasshopper family for whom cannibalism is a concern. The Mormon cricket is another agricultural pest that swarms in large numbers, and Simpson's lab found that they too are driven forward by the desire to cannibalise and avoid being cannibalised; indeed, it was this discovery that inspired Bazazi's work on locusts. This idea puts a new spin on animal migrations as a whole. These mass movements are normally seen as a way for animals to exploit resources that vary across time and space. To migrate successfully, animals are thought to suppress certain routine behaviours like foraging, or indeed cannibalism. But Bazazi's study turns this idea on its head by showing that the threat of cannibalism in itself can drive the movements of migrating hordes.

Surviving open space – the world's hardest animals

In September 2007, a team of scientists launched a squad of tiny animals into space aboard a Russian satellite. Once in orbit, the creatures were shunted into ventilated containers that exposed them to the vacuum of space. In this final frontier, they had no air and they were subjected to extreme dehydration, freezing temperatures, weightlessness and lashings of both cosmic and solar radiation. It is hard to imagine a more inhospitable environment for life but not only did the critters survive, they managed to reproduce on their return to Earth. Meet the planet's toughest animals – the tardigrades.

Tardigrades are small aquatic invertebrates that are also known as "water bears" partially because of their shuffling and unfeasibly cute walks. They also happen to be nigh-invincible and can tolerate extreme environments that would kill almost any other animal. They can take temperatures close to absolute zero, withstand punishing doses of radiation and live through prolonged periods of drought. And now, they have become the only animals to have ever survived the raw vacuum of space. Their stellar adventure began with Ingemar Jonsson from Kristianstad University, who really wanted to test the limits of their resilience. To that end, he launched adults from two species (*Richtersius coronifer* and *Milnesium tardigradum*) into space aboard the FOTON-M3 spacecraft, as part of a mission amusingly known as TARDIS (Tardigrades In Space). After blast-off, they spent ten days in low Earth orbit, about 270km above sea level.

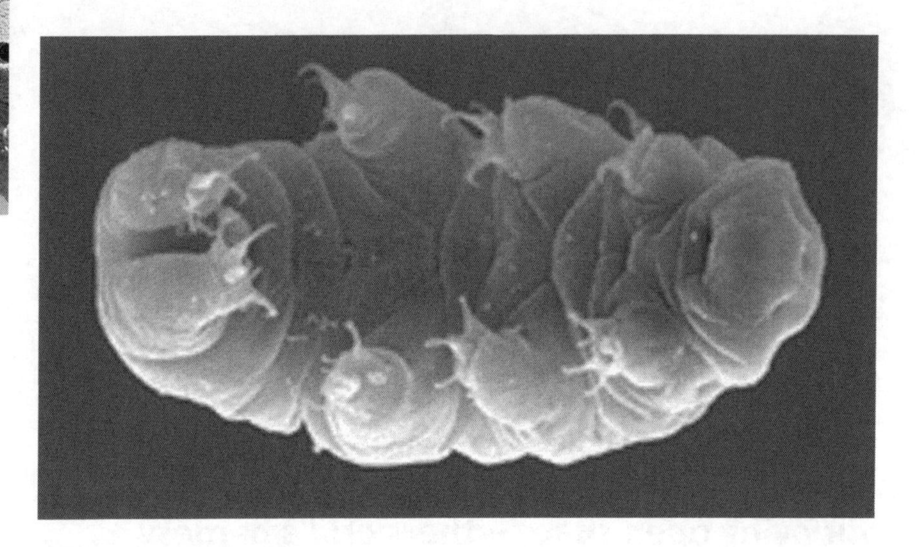

The tardigrades were sent into orbit in a dry, dormant state called a "tun" and it is this desiccated form that is the key to their extraordinary levels of endurance. By replacing almost of the water in their bodies with a sugar called trehalose, they can escape many of the things that would otherwise kill them. Jonsson says, "Environmental agents that rely on water or the respiratory system do not work. You can put a dry tardigrade in pure alcohol and expose them to poisonous gases without killing them." The ability to dry out completely is an adaptation to the tardigrades' precarious environment – damp pools or patches of water on moss or lichen that can easily evaporate. They have evolved to cope with sporadic drought and can stay dormant for years. All it takes to revive them is a drop of water, and that is exactly what happened when the "tardinauts" returned to Earth. [11]

The majority of both species made it through the vacuum of space and its accompanying cosmic radiation, and were just as likely to still be alive as tardigrades that had remained on the planet. They even managed to lay viable eggs that hatched just as well as their planet-bound peers. Even the eggs themselves shrugged off the inhospitable conditions of space. However, Jonsson did find a limit to their endurance – they struggled to cope with a combination of space vacuum and the high doses of ultraviolet radiation given off by the sun. If their containers were unshielded by UV filters, most of them died as the powerful radiation shattered their DNA.

But not all of them. Even faced with these harshest of conditions, a small number of the tardigrades survived. That is an absolutely incredible feat; these animals were subjected to over 7000 kJm^{-2} of UV radiation, about 350-700 times the amount that a Mediterranean sunbather would soak up. How they cope is a mystery, but Jonsson suggests that they must have very efficient means of repairing DNA damage. So far, only lichens and bacteria have survived an unprotected brush with space, and tardigrades are the only animals to join this illustrious super-mile-high-club. But even they have weaknesses. I asked Jonsson how to go about killing one and it turns out that it is surprisingly easy. For all their resistance to cold, radiation and vacuums, they are "very vulnerable to mechanical damage". You could just squash them.

The tardigrades are not unrivalled in the endurance stakes. If one other group of animals could survive a jaunt in open space, it would surely be the bdelloid rotifers (bdelloid is pronounced with a silent 'b'). They are some of the strangest of all animals. Uniquely, they reproduce entirely asexually and have avoided sex for some 80 million years (more on that on page 46). Like tardigrades, bdelloids are also aquatic invertebrates that can enter a dormant, dried-out state and they too have evolved this ability to cope with environments where water is plentiful but can easily evaporate away. Their adaptations have made them the most radiation-resistant animals on the planet, even more so than the hardy tardigrades.

Ionising (high-energy) radiation is bad news for living cells. In the world of comic books, it grants incredible superpowers but in the real world, it damages DNA. This all-important molecule acts as a recipe book for the various parts of a living thing, and ionising radiation can shatter DNA, effectively tearing up the recipe book into small chunks. Absorbed doses of radiation are measured in Grays and ten of these are more than enough to kill a human. In comparison, bdelloids are a hundred times harder. Eugene Gladyshev and Matthew Meselson from Harvard University found that two species shrugged off as much as 1,000 Grays and were still active two weeks after exposure.[12]

At this dose, their egg-laying capacity took a large hit and fell to 10% of previous levels, but even so, they weren't sterilised completely. Their daughters (who are all identical clones of their parents) were similarly unaffected by the radiation. These figures make bdelloids the most radiation-resistant of all animals so far tested. Even other rotifer groups show similar levels of sterilisation at a fifth of the doses tolerated by bdelloids. Even tardigrades have been sterilised by the 500-1,000 Gray doses tolerated by the bdelloids. This resistance is all the more amazing because radiation affects the DNA of rotifers in the same way as other animals – it shreds it. Gladyshev and Meselson measured the size of the remaining pieces in one species, *Adineta vaga*, immediately after being exposed to radiation. They found that a 560 Gray burst broke the animal's genome in over 500 different places, and the 1,000 Gray doses that they contended with so well created over 1,000 double-stranded breaks. The fact that the bdelloids survive and their offspring are fertile is a clear sign that they have an extraordinary ability to repair these breaks, or to protect the proteins that do so. But most places on earth, including the habitats frequented by rotifers, have very low levels of background radiation and without intense sources, there is no impetus for an animal to evolve extreme resistance. How then could it have evolved?

Other species provide a clue. Only bacteria can give the bdelloids a run for their money in the resistance stakes and one in particular, *Deinococcus radiodurans*, has a name that literally means "terrifying berry that withstands radiation". Like the bdelloids, *Deinococcus* can reassemble a genome that has been torn asunder into tiny fragments. In general, bacteria that are resistant to radiation also tend to be resistant to prolonged bouts of dehydration, a connection that the tardigrades also share. It turns out that both drought and radiation pose similar challenges including the production of damaging reactive oxygen molecules and frequent DNA breaks. So Gladyshev and Meselson believe that the ability to shrug off killer doses of radiation is a happy side-effect of adaptations to dry-living. The resistance to drought may have given the bdelloids a competitive edge over parasites, predators and other rotifers that are not so hardy. It may also have ensured their success when they first started to adopt an asexual way of life, by mitigating some of the more harmful side effects of this strategy.

Without the genetic shuffling that accompanies sex, asexual reproduction is often viewed as a poor long-term strategy that leaves a species unable to adapt quickly to new challenges. But some groups have argued that the process of shattering and reconstructing their genome may provide the bdelloids with genetic benefits that compensate for this drawback. The bdelloids repair their broken DNA by using a duplicate piece as a template for copying the lost information at the site of the break. If this template strand contains a gene with a new beneficial mutation, the animal would suddenly have two copies, and a positive change that might otherwise have been genetically overlooked could more easily spread within a population.

Repeatedly breaking and repairing their genomes could also have protected the bdelloids from parasitic DNA. The genomes of all animals are frequently invaded by selfish bits of DNA called transposons that have the ability to cut themselves out of a genome and insert themselves into a new location, all of their own accord. That can be bad news if the transposon jumps into a location that interferes with an important gene. Our genomes are rife with these genetic parasites but scientists have predicted that their spread would go unchecked in the DNA of an asexual creature, to the point where they do so much damage that the species goes extinct. But not the bdelloids; frequently repairing their DNA may give them a route for cutting out these unwanted genetic parasites. In fact, their genomes, far from being riddled with transposons, are remarkably streamlined and unusually free of these parasitic sequences.

Body of jelly, mouth of steel

Imagine that your hand is made of jelly and you have to carve a roast using a knife that has no handle. The bare metal blade would rip through your hypothetical hand as easily as it would through the meat.

It is clearly no easy task and yet, squid have to cope with a very similar challenge every time they eat a meal. The bodies of squid, like those of their relatives the cuttlefish and octopus, are mainly soft and pliant with one major exception. In the centre of their web of tentacles lies a hard, sharp and murderous beak that resembles that of a parrot. It is a tool for killing and dismembering prey.

The large Humboldt squid (*Dosidicus gigas*) uses its beak to sever the spinal cord of fishy prey, paralysing them for easy dining. The beak is two inches long and incredibly hard (a technical term meaning difficult to dent or scratch), stiff (difficult to bend out of shape) and tough (resistant to fractures). This combination of properties makes the beak harder to deform than virtually all known metals and polymers. That is all the more remarkable because unlike most animal teeth or jaws, it contains no minerals or metals. The beak of the Humboldt is made up solely of organic chemicals and manages to be twice as hard and stiff as the most competitive manmade equivalents. By comparison, the mass of muscle that surrounds and connects to the beak is incredibly soft, the equivalent of a jelly hand gripping a bare metal blade. With such mismatched tissues, how does the squid manage to use its killer mouth without tearing the surrounding muscle to shreds?

Ali Miserez from the University of California, Santa Barbara found the answer.[13] The squid's beak is not a uniform structure. Its chemical composition changes gradually along its entire length, so that the sharp, pigmented tip is a hundred times stiffer than the pliant, translucent base which connects to the soft muscle. It is this gradient that blends the mechanical properties of the beak into those of the tissue around it, and allows the squid to tear through the flesh of its prey and not through its own. It is an absolute marvel of precision bio-engineering.

The powerful beak is not the only reason to be wary of the Humboldt squid. It is a fearsome predator also known, for good reason, as the 'jumbo squid' or 'red devil'. Like all squid, it has a large brain, the ability to change colour and excellent vision. On top of these skills, it adds an aggressive temperament, a two-metre long body, 36 sharp hooks in each of its 2,000 suckers, and a penchant for cannibalism. Diver Scott Cassell regularly swims with Humboldts and has had more experiences with them than most. His accounts give a vivid portrayal of their power and temperament. The first time he tried it (and bear in mind that he is a Special Ops veteran), they beat him up, dislocated his arm and dragged him so fast through the water that his eardrum popped. Now, he wears a custom-made suit of fibreglass-and-Kevlar body armour before he goes in the water.

For the past few years, huge numbers of dead Humboldt squids have washed up on the beaches of America's west coast. The reason for these mass deaths is still unknown but Miserez and his Californian colleagues took advantage of the mysterious events to acquire a large supply of fresh beaks for study. They gradually cut sections away from the beak and analysed the mechanical properties and chemical composition of each one. The Humboldt's beak is made of four key ingredients: water; proteins; chitin (the polymer found in insect exoskeletons); and a dark pigment. Miserez found that the soft base is mostly water (70%) and chitin (25%). Towards the tip, the amount of water and chitin fall, the levels of protein and pigment increase, and the beak hardens. The stabbing point contains 60% protein and 20% pigment, which accounts for its dark black-brown colour. When Miserez dissolved all the components away aside from chitin, he found that this polymer forms an intricate network of fibres, just 30 nanometres thick. This lattice gives the beak its shape but not its mechanical properties, for while chitin is itself very stiff, the dissolved chitin-only beak had a uniformly low stiffness throughout its length.

It is the other components that matter, the proteins and pigment that increase in concentration towards the tip. Proteins are made up of amino acids and these were rich in two amino acids in particular – histidine and Dopa. Miserez found that these two amino acids form bonds with each other, creating an extensive network of cross-linked molecules that give the beak both its stiffness and its dark colour. The darkest portions of the beak were 100 times as stiff as the lightest portions. The water is surprisingly crucial too. Without it, the base would be far less soft and mechanically mismatched to the muscle around it. When Miserez freeze-dried the beak sections to remove all their water, the base layers were only half as stiff as the tip. Why the presence of mere water has this effect on the break is not clear, and it is a question for a later study. Miserez also tantalisingly suggests that Dopa-rich proteins tend to repel water, so the presence of Dopa at the beak's tip could drive water away from it and help to set up the chemical gradient that gives it its properties.

Dopa is an incredibly versatile chemical. It acts as a precursor to pigments like melanin, which make our skin darker when we tan. It also acts as the basis for many of our neurotransmitters – chemicals such as dopamine that are used for signalling in the brain. It is abundant in the biological adhesives used by marine animals, like the sticky threads that form the 'beards' of mussels and the cement used by some species of worms to build their lairs. For these reasons, it has drawn the attention of material scientists eager to duplicate the success of nature's building blocks. Its role in crafting the Humboldt's immaculately constructed beak will only serve to build that interest.

Mexican-waving bees

The forests of east Asia are home to giant honeybees. Each one is about an inch in length and together, they can build nests that measure a few metres across. The bees have an aggressive temperament and a reputation for being among the most dangerous of stinging insects. Within mere seconds, they can mobilise a swarm of aggressive defenders to repel marauding birds or mammals. But against wasps, they use a subtler and altogether more surprising defence – they do a Mexican wave.

Wasps, and hornets in particular, are major predators of bees and the largest ones can make even the giant bees look puny. Some invade hives and steal grubs, while others swoop in and pluck loitering bees from the surface of the colony – a technique known as "bee-hawking". The giant bees are particularly vulnerable to this strategy because their nests are open and typically covered in a blanket of workers. They might seem easy pickings for a hungry hornet, but the workers have a trick up their abdomens. When hornets approach, individual workers raise their rear-ends by ninety degrees and shake them in unison. Nearby workers start doing the same and the result is a ripple of booty-shaking that passes over the surface of the hive. The technical term for the behaviour is "shimmering", but it could be described equally well as a "Mexican moon". Either way, it is mesmerizingly beautiful to watch.

For years, naturalists have assumed that shimmering is an anti-hornet measure but Gerald Kastberger from the University of Graz in Austria has provided the first conclusive proof of this. Together with Evelyn Schmelzer and Ilse Kranner, Kastberger video-taped and meticulously analysed the

behaviour of two colonies of giant bees, nesting at a pair of water towers in Nepal.[14] For them, the study is the culmination of 15 years of observing giant bees in the wild, over the course of many expeditions to India and Nepal. Their videos revealed that there were two types of shimmering – small-scale waves where fewer than 10 bees take part, and large-scale ones which repeatedly spread over the entire surface of the nest and involved hundreds of individuals. The smaller variety were a regular occurrence, but the larger waves only ever happened when wasps were around and became stronger and more frequent if the wasps posed a large threat. More and more individuals joined in the nearer the hornets drew to the nest or the faster they flew towards it, resulting in a defensive zone about 50cm around the nest in all directions. Hornets could happily hover outside this zone but if they crossed it, the bees would start waving strongly. And that had a clear effect on the predators. The moment the shimmering began the hornets halted their attacks and fled, and the more bees that took part, the faster the hornets retreated.

The shimmers had different effects depending on their size; small-scale ones confused the hornets and caused them to veer wildly off course while big-scale waves seemed to irritate them and forced them to briefly accelerate. Either way, the hornets missed the part of the colony that they were aiming for. How it works is not clear but Kastberger thinks that the small-scale waves make it difficult for the hornets to focus on any one bee, while the big-scale ones actually threaten them. The wave fronts tend to stay behind the wasps and they might actually drive them away from their original targeted landing site.

Most species of bees avoid wasps and hornets by nesting in nooks and crannies, but nesting out in the open (see above) has forced the giant bees to evolve other defensive strategies. Interestingly, it seems that the bees use different standards for different predators. Fifty centimetres is apparently enough of a distance for hornets but birds warrant action at much greater distances. In one experiment, giant bees released hundreds of defenders when a kite (the bird of prey, not the toy) appeared about 20 metres away.

When it comes to wasps and hornets, the giant bees use a different tactic called "heat-balling" if any invaders actually manage to land on the nest. A gang of bees seizes and envelops the intruder and starts vibrating their wing muscles. The energetic buzzing heat the bees' bodies to 45°C, a temperature that is harmless to them but lethal to wasps. The bees use their wings to literally cook the intruder to death (and some species can even use the same piles to strangle them).

But in many ways, the shimmering Mexican waves are a much better defensive option. For a start, they are energetically cheap – it takes a lot of effort to mob or heat-ball a hornet, but very little to raise your bum in the air for a split-second. Shimmering is also risk-free. By mooning the hornets away, defenders never actually have to engage with foes that are categorically more powerful than they are. The bees even signal to each other to hold formation. After a few ripples, shimmering workers release a chemical called Nasonov pheromone from a gland on their abdomens. The smell is a message to other bees and it says, "Stay together". It stops guard bees from breaking ranks and flying off to tackle the hornets themselves. Mexican waves might seem like an odd defence but there is no denying their effectiveness. Kastberger says that in 30 minutes of observation and hundreds of attempted attacks by the hornets, he has never seen them take a bee from the nest's surface. Only when the bees left for foraging flights did the hornets successfully catch them in mid-air.

Dogs catch yawns

Dogs may be known for their skills at catching sticks, but new research shows that are just as adept at catching our yawns.[15] The result probably comes as no surprise to dog-owners but it is the first time that it is ever been demonstrated under experimental conditions. Yawning is famously contagious – if one person does it, the chances are that someone nearby will start too. A variety of back-boned animals yawn, but until now only three species were known to catch them from each other – humans, chimps and stump-tailed macaques. The new study provides the first evidence that yawns can be contagious in species other than primates, or that one species can catch a yawn from another.

Ramiro Joly-Mascheroni at Birkbeck University discovered the behaviour by watching how 29 different dogs interacted with a stranger. During a series of trials, the unfamiliar experimenter would call the dogs by name, establish eye contact and then either yawn loudly or silently open their mouths. Twenty-one out of the twenty-nine dogs – 72% in total – yawned at least once when the experimenter yawned. But not a single one of them yawned when the experimenter merely opened their mouth. This stark contrast strongly suggests that the dogs were actually 'catching' the human yawns. In fact, the 'infection' rate of 72% is far higher than the rates at which other animals pass yawns to each other. Chimps only catch yawns about 33% of the time, and humans do so about 45-60% of the time. The fact that the dogs did not yawn in response to any old mouth movement ruled out the possibility that they were just reacting to general movements, or to the presence of an unfamiliar human. It is possible that once the dogs had yawned once by chance, the experimenters were displaying subtle cues that encouraged them to do so again, but that would only have increased the total number of yawns and not the *number* of dogs that did so.

Why would a dog catch a human yawn? It is not that the experimenters were just really boring; Joly-Mascheroni suggests that it is more likely because dogs have a high capacity for empathy. Over the course of domestication, we have selectively bred dogs to be good companions and modern domestic canines are well-known for their ability to read subtle cues from their owners. They will follow the pointing of a finger or the direction of a gaze, and some can even imitate the behaviour of their owners. Some studies have also suggested that empathy – the ability to put yourself in someone else's shoes – also underlies the contagiousness of yawning in humans. People who score more highly in tests of empathy are also more susceptible to catching yawns, while autistic children who have problems with empathy are immune to it. Another possibility is that the dogs may even have learned to catch yawns through their experiences with their human owners.

A third and slightly more improbable explanation is that the yawns were a sign of mild tension or stress. Not all animal yawns indicate tiredness – monkeys sometimes yawn during conflicts and it is possible, though perhaps unlikely, that the dogs viewed human yawns as a confrontational signal. This theory seems unlikely given that the control experiment did not trigger any response but it should at least be easy to disprove by repeating the study while taking some physiological measurements of the yawning canines. It will also be interesting to see if dogs catch yawns from each other. If this behaviour is indeed a side effect of a skill that domestic dogs have evolved for reading human signals, it may be that they are more sensitive to human yawns than those of their own species. Alternatively, it could be that the contagiousness of yawns evolved as a way of communicating within social groups and has since transferred to the partnership between man and hound; if that were so, dogs should be more sensitive to each other's yawns. Those are questions for another time. For the moment, I have learned that reading and writing about yawning for a good half-hour makes you very, very tired. I wonder how the researchers actually managed to get anything done...

An eye for secrecy

Eagles may be famous for their vision, but the most incredible eyes of any animal belong to the mantis shrimp. Neither mantises nor shrimps, these small, pugilistic invertebrates are already renowned for their amazingly complex vision. Now, a group of scientists have found that they use a visual system that has never been seen before in another animal, and it allows them to exchange secret messages.

Mantis shrimps are no stranger to world records. They are famous for their powerful forearms, which can throw the fastest punch on the planet. These awesome appendages lie tucked underneath their heads and draw comparisons to the praying mantises from which they get their name. The arms end in either a fiendish spear or a powerful club and the mantis shrimp can use them to deliver some of the most powerful blows in nature. The arm can accelerate through water at up to 10,000 times the force of gravity, creating a pressure wave that boils the water around it, and eventually hits its prey with the force of a rifle bullet. Both crab shells and aquarium glass shatter just as easily under such an onslaught. But as impressive as their arms are, the eyes of a mantis shrimp are even more incredible.

They are mounted on mobile stalks and can move independently of each other. Mantis shrimps can see objects with three different parts of the same eye, giving them 'trinocular vision'. So unlike humans who perceive depth best with two eyes, these animals can do it perfectly well with either one of theirs. Their colour vision far exceeds our too. The middle section of each eye, the midband, consists of six parallel strips. The first four are loaded

with eight different types of light-sensitive cells (photoreceptors), each of which contains pigments that respond to different wavelengths of light. With this array of photoreceptors, the mantis shrimp can see into the infrared and the ultraviolet parts of the spectrum that are invisible to us. They can even use filters to tune each individual photoreceptor according to local light conditions.

The fifth and six rows of the midband are special – they contain photoreceptors that are specialised for detecting polarised light. Normally, light behaves like a wave and it vibrates in every possible direction as it moves along – think about holding a rope attached to a wall and shaking it in every possible direction. Polarised light vibrates in just one direction, as if you were just shaking the rope up and down. While we are normally oblivious to polarised light, it is present in the glare that reflects off water and glass and we use polarising filters in sunglasses and cameras to screen it out. Many animals can detect polarised light, so it is no surprise that mantis shrimps, with their wondrous eyes, can do so too.

But mantis shrimps can also detect a type of light that no other animal can – "circularly polarised" light. Rather than vibrating up and down, circularly polarised light travels in a spiralling beam. If the path of the light spins clockwise, it is described as "right-handed" and if it spins anti-clockwise, it is described as "left-handed". Tsyr-Huei Chiou from the University of Maryland found that the mantis shrimp's eye contains the only known cells in the animal kingdom that can detect circularly polarised light.[16] Our technology can do the same, but the mantis shrimps beat us to it by about 400 million years.

Peer into the photoreceptor of a mantis shrimp and you would see seven cells arranged in a cylinder. These are called rhabdoms, and each one contains thousands of tiny projections called microvilli. In receptors that are sensitive to polarised light, the microvilli are all arranged in one direction. That creates a narrow slit that only light vibrating in a certain plane can pass through. Three of the seven rhabdoms are sensitive to one plane of polarised light and the other four are sensitive to the plane that is perpendicular to it. But there is an eighth rhabdom sitting in top of these seven, and its microvilli create a slit angled at 45 degrees to those created by the seven cells underneath. This eighth cell converts circularly polarised light into its linear version and in a different way depending on whether it is spinning clockwise or anticlockwise. It is this key innovation that allows the animal to see this unique brand of light.

When Chiou recorded the electrical activity of the seven underlying rhabdoms, he found that some were only sensitive to right-handed circularly polarised light, while others only responded to the left-handed variety. So in theory, mantis shrimps can not only detect circularly polarised light, they can also tell which direction it is spinning in. Chiou provided further evidence of this ability by training mantis shrimps to associate either left-handed or right-handed circularly polarised light with a food reward. After the lessons, he gave them a choice between two food containers that reflected circularly polarised light spinning in different directions. As expected, the animals were more likely to choose the container whose reflections matched those that they had been trained to prefer.

How does this unique visual system benefit a mantis shrimp? For a start, water is replete with circularly polarised reflections and being able to see these could help the animals to see their world in a higher contrast. But Chiou found that the parts of the shells of three species of mantis shrimps also reflect circularly polarised light. See, for example, how different the tail of a mantis shrimp looks under a right-handed circular polarising filter and a left-handed one. Tellingly, males and females produce these reflections from different body parts that are they commonly use to signal to one another during courtship. Chiou speculates that amorous mantis shrimps use circularly polarised light as a secret communication channel. Mantis shrimps use linearly polarised light for this purpose too and while many predators cannot see these codes, they are all too visible to cuttlefish, squid and octopus, which prey on mantis shrimps. The animals avoid that risk by using a signalling method that can be detected by their eyes and theirs alone. Chiou also noted that some species, such as the extraordinarily beautiful peacock mantis shrimp, are more sensitive to circularly polarised light than others. Their communications may be so secret that even other mantis shrimps cannot see them.

Smelly humans

On page 20 of this book, I hurt the mighty reputation of the African elephant (*Loxodonta africana*) by writing about their fear of tiny bees. Now, I think it is time to restore their damaged credibility with research that cements their standing among the most intelligent of animals. African elephants, it seems, can tell the difference between human ethnic groups by smell alone, and they react appropriately to the level of threat that these various groups pose.[17] The Massai, for example, are a group of cattle-herders, whose young men sometimes prove themselves by spearing elephants. They are best avoided and it would clearly benefit an elephant to be able to sort out these humans from those who pose little threat, like the Kamba, a group of pastoralist cattle-herders that do not hunt elephants.

At the Amboseli National Park in Kenya, Lucy Bates found that elephants reacted more fearfully to clothes previously worn by a Massai man than to clean ones or those worn by a Kamba man. She placed the three types of cloth near 18 family groups and watched what happened. When the first individual caught whiff of a new scent, it raised its head and curled its trunk towards the source of the smell. If they smelled Massai clothes, they moved away particularly fast, travelled about five times further and took more than twice as long to relax. They could clearly tell the difference between the two groups based on smell and reacted more defensively to the dangerous one. Every single time the elephants smelled Massai on the wind, they moved downwind and did not stop until they reached tall (and aptly named) elephant grass, over a metre in height. By contrast, they only sought the shelter of the tall grass in about half of the trials with Kamba clothes, and almost none of the trials with clean clothes. To Bates, heading for the grass was a clear sign of planned action – elephant grass only covers about 7% of Amboseli so it is very unlikely that the elephants were just stumbling across it by chance.

In all cases, the elephants never approached within 10m of the clothes and would not have seen them. They reacted on smell alone. It is possible that the Massai and Kamba exude different pheromones, but their distinctive scents possibly stem from their vastly contrasting cultures. The Kamba eat meat, vegetables and maize meal. The Massai, on the other hand, subsist mainly on milk and cattle meat and their villages are suffused by the smells of their herds. They also use ochre and sheep fat for decorations. To an elephant's sensitive trunks, the resulting smells must be as different as red and green beacons are to our eyes. Visual cues worked too – while most African groups wear a wide range of colours, the Massai

traditionally wear a striking red. Bates found that when the elephants saw clean, unworn red cloths, they reacted much more aggressively than they did to clean white ones, even though red is a fairly drab colour to elephant eyes. Bates believes that the smell of Massai triggers a strong fear in the elephants that overrides whatever their eyes tell them. If the elephant sees the distinctive colour of the Massai but does not smell them, its lack of fear allows aggression to come to the forefront.

Seven of the elephant families included individuals that had faced spears over the last 30 years, and two individuals in particular had a history of violence towards Massai cattle. But neither of these factors affected the elephants' reactions. Even groups that were personally inexperienced with Massai spears showed similarly strong reactions to the veterans. Like the bee study, this once again shows that elephants rely heavily on shared knowledge, even between different family groups. Elephants groups have complex social structures and they can recognise individuals in the group by their calls. They appear to recognise the bones of dead elephants and mourn members of their group who pass away. Many non-human animals including meerkats, vervet monkeys and prairie dogs can classify predators into different groups and react accordingly to the type of threat they pose. But so far, the elephant's ability to split members of a single species into further sub-groups is a unique one.

Life as an egg

The mayfly is known for its incredibly short adult life. After spending months as larvae, the adults finally hatches only to fly, mate and die within the space of a day. Now, in the dry south-west corner of Madagascar, scientists have discovered the lizard equivalent of the mayfly – Labord's chameleon *(Furcifer labordi)*. The lifespan of Labord's chameleon is hardly as compressed as that of a mayfly, but it is extraordinarily short for a tetrapod (an animal with four legs and a backbone). From laying of egg to kicking of bucket, the lizard's entire life is played out in a year, and seven months of that is spent inside the egg. The adult chameleons hatch in unison in November and in April, the entire population dies *en masse*. We know the lifespans of over 1,700 species of tetrapods and none are as short as the Labord's. In fact, the vast majority of mammals, birds, reptiles and amphibians live for several years, if not decades, and extreme longevity is fairly common. Whales, giant tortoises, some parrots and indeed, some humans only pop their clogs after more than a century of life. In contrast, very few tetrapods have adopted strategies at the other extreme, where life involves a rapid race to maturity and death occurs within less than a year. Until now, the only tetrapods known to do so were a handful of marsupial mice and opossums, and even then, only among males.

Kristopher Karsten from Oklahoma State University has changed all that by spending four years studying Labord's chameleon in Madagascar's dry south-west corner.[18] In Madagascar, the wet season begins in November as tropical storms sweep in from the Indian Ocean, and it is then that the first chameleon hatchlings emerge. Most share the same birthday and mature at the same pace, which means that during these months, every single living Labord's chameleon is the same age. The lizards grow quickly, packing on about 2-4% of their body mass every *day*. By early January, they are sexually mature adults and by February, females start to lay eggs, just as the wet season draws to a close. This brief window, when both adults and eggs co-exist is the only point in the year when two generations of Labord's chameleon can be found on Madagascar. After their eggs are laid, the adults' health rapidly worsens, they lose weight, their grip weakens and Karsten saw many of them falling from the trees. By April, all the adults are dead. The eggs remain in a state of arrested development for most of their eight months of incubation, until the arrival of the rains in the following November triggers another round of hatching.

No other tetrapod has a life cycle quite this short, and no other spends such a comparatively large amount of time in the egg. It is unclear why this species in particular has evolved in such an extreme way, but Karsten suggests that Madagascar's harsh and highly seasonal environment may have been a contributing factor. In response to these unpredictable conditions, Labord's chameleon appears to have compressed the majority of its life into a much more stable environment – its own egg. By all accounts, Labord's chameleons live brutal and difficult adult lives. Even their sex lives are harsh, with males competing violently and intensely for mates, and sex itself being fairly aggressive. In general, species that run high risks of being killed as adults tend to grow quickly, mature early and die young – Labord's chameleon is clearly no exception.

The fact that the unusual life cycle of this chameleon has only just been discovered shows how little we know about this apparently familiar group of lizards. There is good reason for that – the majority of chameleons live in Madagascar where they are difficult to find, not least because of their vaunted camouflage skills. For the moment, Karsten's findings have direct implications for conservationists. Chameleons are notorious for dying rapidly in captivity, and this work suggests that this mortality might simply represent a very short, but entirely natural, adult lifespan.

The smallest of heavyweights

Imagine being able to drink ludicrous amounts of alcohol without getting drunk and without the nasty consequences in the morning. For some people, it would be a dream come true but for the pen-tailed tree shrew (*Ptilocercus lowii*), it is just part of everyday life. The tree shrew lives in the rainforests of Malaysia and its local drinking establishment is a large plant called the bertam palm (*Eugeissonia tristis*). The palm develops large stems a few metres in length, each of which sprouts about a thousand flowers. These are loaded with an alcoholic nectar with a maximum alcohol concentration of 3.8% – as strong as beer and one of the most alcoholic natural foodstuffs known. And because of it, the bertam plant regularly smells a lot like a brewery.

Frank Wiens from the University of Bayreuth, Germany, put several palms under 24-hour video surveillance and found that seven species of mammals, including tree shrews, regularly visited them to drink their boozy excretions. They would move up and down the stems for hours at a time, licking nectar as they went. Many animals, from monkeys to elephants, sporadically get hammered on naturally occurring alcohol but aside from some humans, the tree shrew is one of the only species we know of that chugs the demon drink on a daily basis.[19] Wiens calculated that the animals were frequently drinking as much alcohol for their weight as a human woman knocking back nine glasses of wine a day. The little binge-drinkers imbibed these substantial amounts about one night in every three. Wiens confirmed this by testing the hair of captured animals for a chemical called ethyl glucuronide (EtG). Enzymes in the body convert alcohol into EtG, which often becomes incorporated into shafts of growing hair, so that its presence in hair is a sign of previous indiscretions. Sure enough, the tree shrews had far higher levels of EtG in their fur than tee-total lab rats. These levels were so high that if you recorded them from human hair, you would think the person in question was drinking at a life-threatening level.

But amazingly, the tree shrews never showed any signs of drunkenness or failing coordination. Clearly, they must have evolved some way of coping with high doses of alcohol that completely surpasses the abilities of humans. For the moment, we have no idea how they do this, but the high levels of EtG in their hair suggests that they may break down alcohol in this way to a far greater extent than humans do. By converting the alcohol they drink into EtG, they keep the levels in their blood and their brain low.

Wiens believes that the alcoholic nectar is the palm's way of attracting mammals as pollinators. Their flowers are unlike any pollinated by insects, and their sturdy structures and strong-smelling nectar seem tailored to luring small mammals as pollinators of choice, and supporting their weight. Certainly, many of the palm's characteristics seem geared towards brewing its enticing cocktail. The flowers go through an unusually long period of nectar production that gives it enough time to ferment. Their buds act as isolated fermenting chambers and always contained several species of yeast, many of which are new to science.

The relationship between the palm and the tree shrew is an ancient one, possibly dating back to over 55 million years ago. Both species diverged from their respective lineages at about this time and Wiens thinks that they have retained a partnership ever since. The plant gets a pollinator but it is unclear what rewards the tree shrews reap from the alliance. Wiens suggests that drinking alcoholic nectar provides them with some sort of advantage, which balanced out the drawbacks of chronic drinking long enough for the animals to evolve ways of coping. The fact that the palm churns out their nectar at regular intervals throughout the year may have something to do with it – it makes them a dependable supply of food.

The pen-tailed tree shrew is the most primitive of the tree shrews, which are themselves an early offshoot of the lineage that led to monkeys, apes and eventually us. As such, they could prove to be an interesting model for understanding the role of alcohol in the evolution of the primates.

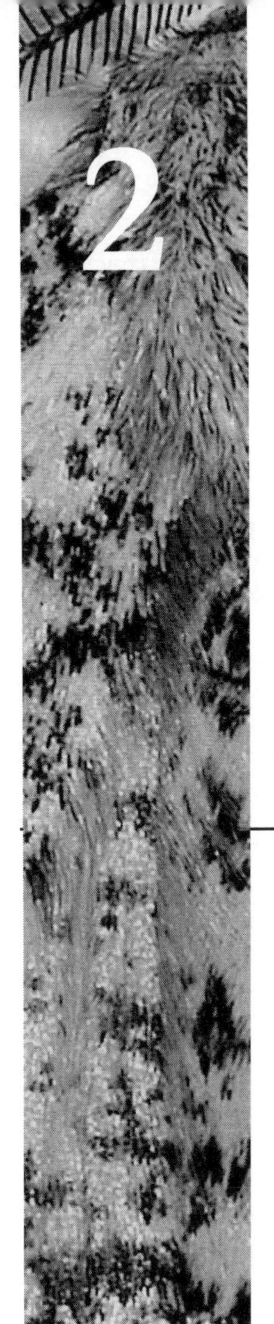

A tangled bank

Picasso fish, hoofed proto-whales, evolving verbs, and how to replay the tape of life

How to see on your side

Imagine watching a movie where every now and then, key frames have been cut out. The film seems stilted and disjointed and you have to rely on logic to fill the gaps in the plots. Evolutionary biologists face a similar obstacle when trying to piece together how living species evolved from their common ancestors. It is like watching a film with a bare minimum of footage. The species alive today represent just a few frames at the very end, and the fossil record is a smattering of moments throughout the film's length. But the gaps, while plentiful, are being slowly filled in. With amazing regularity, new fossils are being unearthed that bridge the gap between existing specimens. These "transitional fossils" are always greeted with great relish for their intermediate nature provides yet more examples of gradual evolution from one form to another. They act as handy visual aids for explaining the story of evolution to those with a dearth of imagination.

This year, Matt Friedman from the University of Chicago described a new transitional fossil that is one of the most dramatic yet.[20] Its name is *Heteronectes* (meaning "different swimmer") and it is a flatfish, but not as you know it. You have probably eaten flatfish before but tasty fillets of plaice, sole or halibut give few hints about their extraordinary physical specialisations. They are fish that live on their sides and their flat profiles and skill at concealment make them both efficient hunters and difficult prey. For other fish, lying sideways would give one eye a useless view of sand but flatfish have adapted accordingly. Their young resemble those of other fish but as they grow, one of their eyes makes an amazing journey to the other side of its head. The adults look like they have swum out of a Picasso painting. These bizarre faces evolved from the standard symmetrical eye positions that almost all other fish have. But *Heteronectes* is a half-committed flatfish. Like modern representatives, its skull was asymmetrical but while one eye had begun migrating to the other side of its head, it stopped near the midline without crossing over fully. No living flatfish has eyes arranged in such a way. We could not have wished for a better intermediate form – it is a marvellous missing link between the standard fish body plan and the distorted visages of flounders and soles.

The discovery of a transitional flatfish is particularly poignant because the group's grotesque faces have fuelled a significant amount of debate. In *The Blind Watchmaker*, Richard Dawkins views the displaced eye as a perfect example of adaptation, an extraordinary solution to the problems posed by life on your side. But opponents of evolution have long viewed the flatfish's shifted face as proof against gradual adaptation, for what would be the

point of a half-migrated eye that would still be facing into the sand? Darwin himself was somewhat stymied by flatfish, and proposed an explanation with shades of Lamarckism – the mostly discredited idea that organisms can inherit traits that their parents acquired during their lifetimes. As an alternative, Robert Goldschmidt claimed that the flatfish's eye provided evidence for his "Hopeful Monster" hypothesis, which suggested that some features evolved extremely suddenly, bypassing the need for any intermediary forms. But *Heteronectes* quashes both these hypotheses and confirms that the deformed body plan of modern flatfish developed with a gradual, step-wise tempo.

Friedman also studied another intermediate flatfish called *Amphistium* which hailed from the same region of northern Italy. Palaeontologists had previously dismissed *Amphistium*'s resemblance to modern flatfish but by bringing modern scanning technology to the fore, Friedman has confirmed its membership within the group. What's more, *Amphistium*'s asymmetrical skull, like that of *Heteronectes*, bears an eye that has only migrated halfway across its head. Could these fossils simply be juvenile flatfish whose eyes have not finished their migration? Not according to Friedman, who offers three lines of evidence that say otherwise. In living flatfish, the eye starts to migrate when the fry hit a centimetre or so in length, and all the known specimens of *Amphistium* and *Heteronectes* are more than ten times as long. Their skulls have completely hardened, which only happens in adult flatfish once their eyes have stopped moving. And even though specimens of *Amphistium* have a wide range of lengths, all of their eyes have the same alignment. The fossils are the remains of adults.

Nor is it likely that *Amphistium* and *Heteronectes* belong to different groups of fish whose skulls have been crushed and distorted. In the *Heteronectes* fossil, the areas around the eyes are obviously asymmetrical but they show no signs of twisting damage, and other parts of the skull have not been deformed. Their fins and tails bear features that are trademarks of the flatfish dynasty but some of these are only found in the group's most primitive member – the spiny turbot. The spiny turbot shares another trait with *Amphistium* – both species are ambivalent in their asymmetry and have both right-sided and left-sided individuals. Other modern flatfish have a dominant side, and that strongly suggests that *Amphistium* and *Heteronectes* are not advanced flatfish whose eyes happen to have reverted to a state of incomplete migration. They are indeed ancestral members of the group and they sit outside the lineage that includes modern species.

In the light of such dramatic fossils, asking about the point of half-migrated eyes is not going to cause advocates of evolution to lose much sleep. It is intuitive to suggest that such organs would be useless, but the point is moot when they clearly existed! Still, it is an interesting question – how did *Heteronectes* cope with an eye that, while displaced, would still have had an eyeful of sand? Friedman suggests that *Heteronectes* could have propped itself up by using its dorsal fin, so that its head was lifted just high enough above the surface to give it a view. That is a speculative guess, but it is not a wild one – modern flatfish can perform the same trick, and both *Heteronectes* and *Amphistium* had even longer bony rays in their dorsal fins. And what initially compelled a vertically flattened fish to start lying on its side? The prevailing theory is that it provided them with a perfect posture for springing an ambush. It is a hypothesis that neither *Heteronectes* nor *Amphistium* is in a position to confirm or deny. But at the very least, it is clear that *Amphistium* was a hunter of other fish, for one fossil contains the remains of a smaller fish inside its stomach. Perhaps other missing links will provide even more answers.

The (jellyfish) eyes have it

Jellyfish may seem like simple blobs of goo, but some are surprisingly sophisticated. The box jellyfish (*Tripedelia cystophora*), for example, is a fast and active hunter and stalks its prey with the aid of 24 fully functioning eyes. These are grouped into four clusters called rhopalia, which lie on each side of its cube-like body. Together, they give the box jellyfish a complete 360 degree view of its world and make it highly manoeuvrable. Each eye cluster contains six eyes. Four of these are merely simple pits containing light-sensitive pigments, but two are remarkably advanced and carry their own lenses, retinas and corneas. The lenses are good enough to produce images that are free of distortion and even though the views are blurrier than those we see, these complex 'camera-type' eyes are very similar to those of more advanced animals like ourselves and other vertebrates.

These similarities extend to a more fundamental level. Even though jellyfish are the most ancient group of animals to have a well developed visual system, it turns out that their eyes are built with many of the same genetic building blocks that ours are. All animal eyes, from the familiar human version to the compound eye of insects, contain two basic components. They have a photoreceptor – a cell that converts streams of light into chemical signals – and a dark pigment that focuses these streams. The photoreceptors share a basic plan too – they always work through a partnership between a protein called an opsin and a chemical called retinal. When light strikes retinal, the molecule's shape changes and it separates from opsin. That triggers a chemical signal that ends in an electrical impulse travelling to the brain.

Vertebrates and invertebrates differ in both the pigments and the photoreceptors they use, and both groups have their own distinctive opsins and signalling cascades. Zbynek Kozmik from the Academy of Sciences of the Czech Republic found that the box jellyfish is unusual in the structure of its photoreceptors, which are closer to those of the back-boned vertebrates than the spineless invertebrates.[21] When Kozmik looked for box jellyfish genes that are involved in sight, he found that their opsin protein is also similar to the versions found in vertebrate eyes. With further tests, Kozmik confirmed that the jellyfish's opsin is a fully functioning visual protein. It sticks to retinal and is particularly sensitive to blue-green light. The similarities did not stop there. Kozmik found that the chains of proteins that carry the message passed on by opsin are again similar to those used by vertebrates. And just as our eyes use the dark pigment melanin, so do those of the box jellyfish. Amid its genome, Kozmik found the jellyfish versions of human gene called *oca2* and *mitf* that are essential for creating melanin. The genes are switched on in a part of the jellyfish eye that is littered with granules of pigment, which were identified as melanin through chemical tests.

Despite the massive evolutionary gulf that separates jellyfish and vertebrates, both groups construct their eyes using similar genetic components. It is possible that they kept an ancient 'eye program' that their shared ancestor already had, but Kozmik thinks that this is unlikely. If any such program existed, it must have eventually been abandoned by many animal groups, for most sighted invertebrates, such as octopuses and insects, build their eyes with a very different set of genes. Kozmik argues that eyes provide such an important advantage that there is no obvious reason why any group of animal should abandon one working system of building them, in favour of a completely different one. Instead, it is more likely that jellyfish and vertebrates evolved their eyes by independently recruiting the same genetic building blocks, in a case of parallel evolution. That is not unfeasible; there are other examples of large networks of genes being co-opted for new purposes, and computer models have estimated that it would only take about half a million generations to evolve a sophisticated camera-type eye from a simple patch of light-sensitive cells – barely a blink in evolutionary time. In fact, it is likely that the jellyfish's advanced camera-type eye evolved from the primitive cup-like versions that sit next to them on the rhopalia. These simpler eyes contain proteins called crystallins, which help to build the lenses of the advanced ones. And the *mitf* gene which helps to produce melanin in the camera-type eyes is also active in the cup-like ones. The eyes of the box jellyfish tell us yet again that important innovations, such as eyes, evolved by changing how existing groups of genes are used, rather than adding new ones to the mix.

Hoofed proto-whales

Travel back in time to about 50 million years ago and you might catch a glimpse of a small, unassuming animal walking on slender legs tipped with hooves, by the rivers of southern Asia. It feeds on land but when it picks up signs of danger, it readily takes to the water and wades to safety. The animal is called *Indohyus* (literally "India's pig") and though it may not look like it, it is the earliest known relative of today's whales and dolphins. Known mostly through a few fossil teeth, a more complete skeleton was described for the first time last year by Hans Thewissen and colleagues from the Northeastern Ohio Universities.[22] It shows what the missing link between whales and their deer-like ancestors might have looked like and how it probably behaved.

Whales look so unlike other mammals that it is hard to imagine the type of creature that they evolved from. Once they took to the water, their evolutionary journey is fairly clear. A series of incredible fossils have documented their transformation into the masterful swimmers of today's oceans from early four-legged forms like *Pakicetus* and *Ambulocetus* (also discovered by Thewissen). But what did their ancestors look like when they still lived on land? Until now, we had little idea and their modern relatives have provided few clues. According to molecular evidence, the closest living relatives of whales are, quite surprisingly, the artiodactyls, a group of hoofed

mammals that includes deer, cows, sheep, pigs, giraffes, camels and hippos. They all have a characteristic even number of toes on each hoof and not a single one of them bears even a passing resemblance to whales and dolphins. Among the group, the hippos are evolutionarily closest and while they are at least at home in water, their family originated some 35 million years after the first whales and dolphins did.

Enter *Indohyus*, a small animal about 70cm long that lived 47 million years ago. It was a member of a family of mammals called the raoellids, prehistoric artiodactyls that lived at the same time as the earliest whales and that hailed from the same place of origin – southern Asia. By analysing a fossilised skull and a set of limbs collected from India, Thewissen found compelling evidence that the raoellids were a sister group to the ancestors of whales. Even though *Indohyus* had the elegant legs of a small deer and walked around on hooves, it also had features found only in modern and fossil whales. Its jaws and teeth were similar to those of early whales, but the best evidence was the presence of a thickened knob of bone in its middle ear, called an involucrum. This structure helps modern whales to hear underwater; it is only found in whales and their ancestors, and acts as a diagnostic feature for the group – all whales have an involucrum and all living animals with an involucrum are whales. Based on these physical similarities, Thewissen suggests that the raoellids are a sister group to the whales. Both of these groups are evolutionary cousins to all modern artiodactyls. (Note that Indohyus is *not* a direct ancestor of whales, as many news reports claimed at the time. Nor did whales "evolve from deer"!)

Indohyus's skeleton also suggests that it was partially adapted for life in the water. Its leg bones were unusually thick, a feature shared by other aquatic animals including hippos, sea otters and manatees. These heavier bones stop swimming mammals from floating by default and allow them to hang in the water and dive more easily. Because *Indohyus* had slender legs and not paddle-shaped ones, Thewissen pictures it wading in shallow water, walking hippo-style along the river floor while its heavy bones provided ballast. Thewissen found more clues about the animal's lifestyle from its teeth, and particularly the levels of certain isotopes in their enamel. Levels of oxygen isotopes matched those of water-going mammals, providing further support for *Indohyus*'s aquatic tendencies. Its large crushing molars are typical of plant-eaters and levels of carbon isotopes in them suggested that *Indohyus* either came onto land to graze (like hippos) or fed on plants and invertebrates in the water (like muskrats). In terms of behaviour, they were close to the modern mouse-deer, a tiny, secretive deer that feeds on land but flees into streams when danger threatens.

Put together, this portrait of *Indohyus*'s life also tells us about the changes that drove the evolution of whales, and it looks like it was not a move to water. Whales and raoellids are evolutionary sisters and since early members of both groups were happy in the water, aquatic lifestyles must have pre-dated the origin of whales. Instead, Thewissen suggests that the key step was a switch in diet. He speculates that whales developed from an ancestor that was similar to *Indohyus*, which fed on plants and possibly small invertebrates on land, but fled to water to escape predators. Over time, they slowly turned into meat-eaters and evolved to swim after nimble aquatic prey.

80 million years without sex

Sex is, on the whole, a good thing. I know it, you know it, and natural selection knows it. But try telling it to bdelloid rotifers. These small invertebrates have survived without sex for some 80 million years. While many animals, from aphids to Komodo dragons, can reproduce asexually from time to time, it is incredibly rare to find a group that have abandoned sex altogether. The bdelloid rotifers (pronounced with a silent b) are an exception. They live in an all-female world and since their discovery, not a single male has ever been found. Genetic studies have confirmed that they are permanently asexual, and females reproduce by spawning clone daughters that are genetically identical to them. The bdelloids are no strangers to attention-grabbing science (see page 23 for more on their extreme survival skills) but they pose a problem for evolutionary biologists, who have struggled to explain how they could make do without a strategy that serves the rest of the animal kingdom very well.

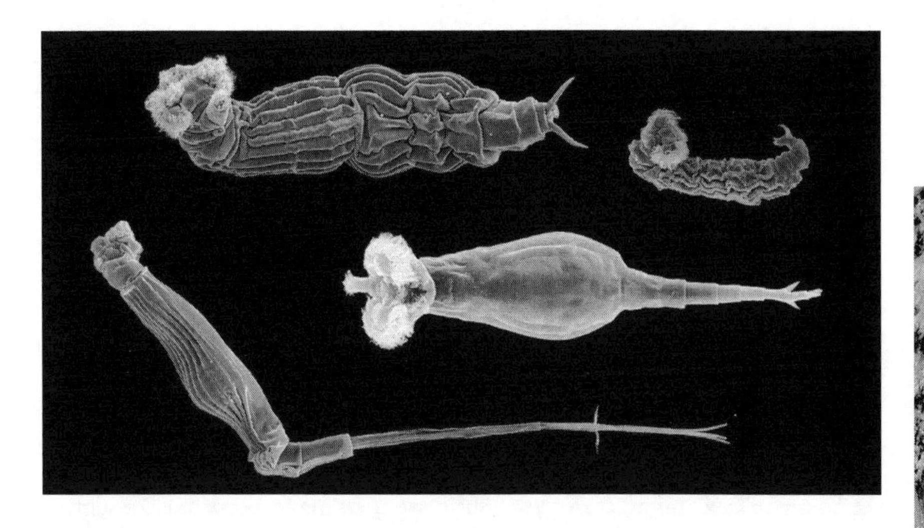

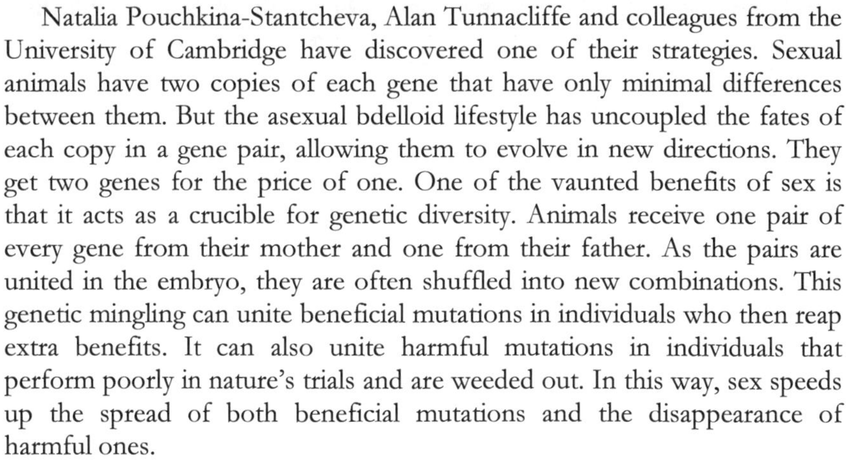

Natalia Pouchkina-Stantcheva, Alan Tunnacliffe and colleagues from the University of Cambridge have discovered one of their strategies. Sexual animals have two copies of each gene that have only minimal differences between them. But the asexual bdelloid lifestyle has uncoupled the fates of each copy in a gene pair, allowing them to evolve in new directions. They get two genes for the price of one. One of the vaunted benefits of sex is that it acts as a crucible for genetic diversity. Animals receive one pair of every gene from their mother and one from their father. As the pairs are united in the embryo, they are often shuffled into new combinations. This genetic mingling can unite beneficial mutations in individuals who then reap extra benefits. It can also unite harmful mutations in individuals that perform poorly in nature's trials and are weeded out. In this way, sex speeds up the spread of both beneficial mutations and the disappearance of harmful ones.

Asexual reproduction carries none of these benefits and some scientists see it as a poor and plodding long-term strategy that leaves a species unable to quickly adapt to new challenges. But the bdelloids are proof that permanent asexuality works and they have a way of creating genetic diversity that is all their own. The genetic shuffling that accompanies sex means that even though an individual has two copies of every gene, they are mostly the same. Small differences between the two copies can have significant effects, but they are small differences nonetheless. For example, in other rotifer groups that still practice sex, an individual's gene pairs are at least 97.6% identical to each other. But in the asexual bdelloids, a daughter inherits both copies of each gene from its mother and they never mingle. This frees them to evolve in their own directions and take on new roles, irrespective of the destinies of their partner.

Pouchkina-Stantcheva found evidence of this in a gene called lea from the bdelloid *Adineta ricciae*.[23] The two copies of this gene differ by about 14% of their sequence, and these small differences translate to proteins with substantially different structures and functions. Bdelloids live in freshwater pools but can survive periods of drought by dehydrating and living in a dry and dormant state. Both LEA proteins protect the animal in this state, but they do so in complementary ways. Version A acts as a molecular shield and prevents more sensitive proteins from balling together into useless clumps when they dry out. Version B insinuates itself into the fatty membranes that surround all cells and helps to stabilise them. By sending two copies of a single gene down different evolutionary paths, natural selection has provided *A.ricciae* with two separate ways of protecting itself from drought. The researchers believe that this two-for-the-price-of-one deal on protein diversity helps to compensate for the loss of diversity that accompanies a shunning of sex. Many major evolutionary steps have occurred through a similar process called 'gene duplication'. During the genetic shuffle that accompanies sex, a gene is mistakenly duplicated and the extra redundant copy is free to evolve new functions. The bdelloids use essentially the same process but they do not need to create another copy of their genes when they already have two independent ones.

Nor do they have to simply contend with the genes from their own genomes; they can import some from other species. Their genomes are rife with legions of foreign DNA, transferred from bacteria, fungi and even plants. The swapping of genetic material is all part of a day's activity for bacteria but it is incredibly rare in animals. Bdelloids, however, bring in external genes to an extent that is completely unheard of in complex organisms. Each rotifer is a genetic mosaic, whose DNA spans almost all the major kingdoms of life. Eugene Gladyshev and colleagues from Harvard University (who also discovered the bdelloids' incredible resistance to radiation) came across their genetic smuggling almost by accident. Gladyshev initially set out to study a group of DNA sequences called transposons, which have the ability to jump around genomes. But among the DNA of the bdelloid *Adineta vaga*, he unexpectedly found traces of bacterial, fungal and even plant genes, many of which are incredibly rare.[24] About half have no counterparts in animals and one gene is found in only 10 species of bacteria. Some genes appear to have been smuggled into the rotifer genome as a set, for they appear in the same order and orientation that they do in fungi or bacteria. The genes are not passive hitchhikers either; many appear to be fully functional. In their native species, they are involved in breaking down sugars and carbohydrates, or producing useful molecules like antibiotics and toxins. The chances are that the rotifers are putting them to similar uses. The recruitment process is also an ongoing one. Some of the foreign genes were clearly brought in so long ago that they have picked up characteristics that are similar to the rotifer's own genes. Others, which are relatively unchanged from their counterparts in bacteria or fungi, are probably more recent additions.

It is not clear why the bdelloids are so good at incorporating new genes, but their lifestyle may hold the answer. The freshwater ponds they call home frequently dry out and they cope with this by entering into a dry, dormant and extremely tough state. This drying process breaks their cell membranes and shatters their DNA. The bdelloids are very good at repairing these injuries, but it provides a temporary entry point for chunks of foreign DNA from species in the surrounding environment that have also succumbed to similar damage. This newly discovered ability may help to explain the success of the bdelloids despite their rejection of sex. Compared to sex, asexual reproduction is often seen as a poor long-term strategy but bdelloids have contradicted this theory by being very successful; there are over 360 species alive today. Gladyshev suggests that this success may be due to their ability to pick up new genes from their environment. If the main advantage of sex is that it promotes genetic diversity, why worry about it when you have the gene pools of entire kingdoms available to you?

Replaying life's tape

When it comes to evolution, a species' past has a massive bearing on what it might become. That is the latest message from a 20-year-long experiment in evolution, which shows how small twists of fate can take organisms down very different evolutionary paths.

The role of history in evolution is a hotly debated topic. The late Stephen Jay Gould was a firm believer in its importance and held the view that innocuous historical events can have massive repercussions, often making the difference between survival and extinction. To him, every genetic change is an "accident of history" that makes some subsequent changes more likely and others less so. Evolution, as a result, is "fundamentally quirky and unpredictable". In his book *Wonderful Life*, Gould imagined that if we replayed life's tape from some point in the past, evolution would go down very different paths than the ones it has currently taken. Another eminent palaeontologist, Simon Conway Morris, disagreed. He argued that life can weave its way down any number of evolutionary routes, but that its "destinations are limited". He saw the fact that living things often converge on the same adaptations as evidence that history has very little pull on evolution. In his mind, replaying life's tape would lead to more or less the same result, with historical contingency only altering minor details.

Of course, it is impossible to replay life's tape on a planetary scale but some experiments allow us the chance to do so at a more modest level. The aptly named "long-term evolution experiment" (or LTEE) is one of these. It is the longest-running experiment in evolution ever undertaken and began in 1988, when Richard Lenski at Michigan State University bred 12 lines of the gut bacteria *Escherichia coli* from a single ancestor. Since then, the bacteria have been grown in twelve separate vials of sugary broth and plopped into fresh solution every day. Every 500 generations, samples are frozen to act as a 'fossil record', and since the experiment's humble beginnings, over 44,000 generations have passed. In this time, the bacteria have changed greatly. All of them are now bigger, grow faster on sugar, and take less time to establish new colonies. But recently, Lenski noticed that one lineage of bacteria have developed an extremely rare adaptation that has turned up only once in the entire history of the experiment.[25] Why?

The bacteria are grown in broth that is low in sugar, and they usually run out by the afternoon, when they are forced to stop growing. The broth is also rich in citrate ions but in general, *E.coli* cannot use these as fuel when oxygen is around. Any bacterium that evolves this ability would suddenly find itself amid a vast and exclusive energy source. But in 20 years of evolution, only one population of *E.coli* has managed to do this, even though all 12 strains have evolved under exactly the same conditions and all 12 have been exposed to citrate from the beginning. Blount first discovered the citrate adaptation, when he noticed that one vial of bacteria (known as Ara-3) was much cloudier than usual, a sign that the cells that been growing on a fuel other than sugar. Unlike almost all other *E.coli*, some of the cells from Ara-3 were able to grow solely on citrate. To prove that these unique cells did not come from a contaminating strain, Blount showed that their DNA carried certain giveaway mutations that characterised earlier generations of Ara-3.

During the experiment, each population of bacteria experienced billions of mutations, and given the size of *E.coli*'s genome, it is likely that each lineage tried every typical genetic change several times. And yet, the citrate adaptation turned up only once in all of this tinkering. Based on the experiment, the odds of a bacterium developing the adaptation in any generation is just one in ten trillion! There are two possible explanations for this: it is the result of an extremely rare mutation; or it depended on the particular history of that specific population. To work out which, Zachary Blount from Lenski's lab used the frozen fossil record to replay evolution

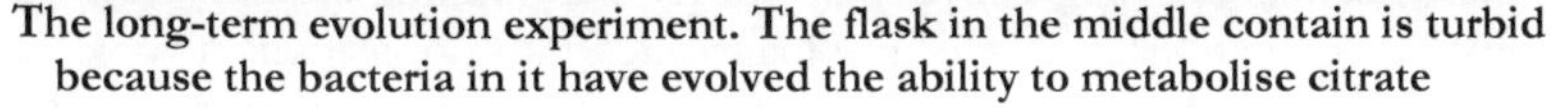

The long-term evolution experiment. The flask in the middle contain is turbid because the bacteria in it have evolved the ability to metabolise citrate

from different points from this population's past. In three separate replays, Blount cultured bacteria from the entire length of the experiment, from the original ancestor to the most recent generation. In all cases, he found that citrate users evolved much more often in samples from later generations than those from earlier ones. These results strongly suggest that the citrate innovation was not simply the consequence of an extremely rare mutation. If that was the case, it would be equally likely to crop up at any point throughout the experiment's history. To Blount, the fact that it happened more frequently in later generations suggests that it depended on one or more mutations that the bacteria had previously picked up, which unlocked the potential for the later innovation.

It is possible that the ability to use citrate depends on teamwork between several genes. Any one of these could pick up the right mutation, but it would be meaningless if earlier mutations had not provided the right partners. Alternatively, earlier mutations could have been directly responsible for later ones. Blount speculates that the clinching mutation might have happened in a piece of mobile DNA that had previously jumped into the right spot. Without this insertion, the critical change would never have happened. Blount and Lenski are now trying to discover this sequence of genetic events, and how they affected the bacterial cells. Whatever the route, one lineage of bacteria had managed to make use of citrate after about 31,000 generations. Their numbers grew quickly and by 32,500 generations, they made up 20% of the population. But 500 generations later, they had fallen back down to just 1%.

Blount believes that the first bacteria to use citrate just weren't very good at it and reaped only marginal benefits from their innovation. As a result, they initially prospered but were soon outcompeted by other bacteria that could not use citrate but had become much better at metabolising sugar. Only later did they regain their initial foothold, with further mutations that improved their unique ability and allowed them to switch seamlessly from sugar to citrate when the first fuel ran out. Eventually, these advances catapulted the citrate users to dominance. Their sugar-only peers still eke out a minority existence because they are still better at using sugar as an energy source. So a single innovation – citrate exploitation – was enough to split a united population of bacteria into a community of two members: a specialist that focuses on sugar; and a generalist that uses both sugar and citrate. The team describes the implications of the experiment thus:

> "Historical contingency can have a profound and lasting impact under the simplest conditions, in which initially identical populations evolve in identical environments. Even from so simple a beginning, small happenstances of history may lead populations along different evolutionary paths. A potentiated cell took the one less travelled by, and that has made all the difference."

Flight before echoes

The heads and bodies of bats have amassed an extraordinary array of adaptations that have make them lords of the night sky. Today, the thousand-plus types of bats make up a fifth of living mammal species. Richard Dawkins once described the evolution of bats as "one of the most enthralling stories in all natural history". Now, we are in a good position to work out how that story began.

The success of bats hinges on two key abilities: their mastery of flight, a feat matched only by birds and insects; and echolocation, the ability to navigate their way through pitch-blackness by timing the reflections of high-pitched squeaks. For evolutionary scientists, the big question has always been: which came first? Until now, fossil bats have not provided any clues for all of them show signs of both echolocation *and* flight. But a stunning new fossil, discovered by Nancy Simmons from the American Museum of Natural History is an exception and it provides a categorical answer to the long-running debate – the earliest bats could fly but could not echolocate.

The new creature hails from the Green River in Wyoming and is known as *Onychonycteris*, meaning "clawed bat".[26] Its fossils date back to about 52.5 million years ago and by comparing it to other prehistoric bats, Simmons found that it is the most ancient member of this lineage so far discovered. It acts as a 'missing link' in bat evolution, much like the famous *Archaeopteryx*, which hinted that birds may have evolved from dinosaurs. *Onychonycteris* was clearly a capable flyer. Its wings are remarkably similar to those of today's bats, with the exception of small claws at the end of its digits that modern bats lack. But the really exciting part of the new fossil is the size of its cochlea – the coiled tube that allows mammals to hear. Relative to the size of their head, all bats that use echolocation have massively enlarged cochleae and the two traits are so tightly linked that large cochleae have been used to show that other prehistoric bats were also echo-locators. Not so for *Onychonycteris*. The dimensions of its skull revealed that its cochlea was too small to have supported echolocation and was closer in size to those of flying foxes, large bats that favour vision over echolocation. It provides direct evidence that bats mastered the art of flying before they developed a way of navigating through the dark.

When the flight-echolocation debate first started, the "echolocation-first" camp had the upper hand. According to this school of thought, the earliest bats used echolocation from tree perches to detect and snatch passing insects, and indeed, the most primitive of modern bats sometimes use this technique. The ancestral bats then evolved long, webbed arms and fingers to better catch their prey and eventually took to jumping after them. After leaping, their long, webbed fingers would have allowed them to glide to safety and eventually developed to the point where flapping flight was possible. However, this theory has since run into troubled times. It turns out that echolocation – which involves producing very loud bursts of sound – takes a lot of energy and it is something that a perching bat cannot afford to keep up for long. A flying bat on the other hand, has no such difficulties. Every time they flap their wings, bats contract the powerful lung muscles that power their ultrasonic shouts; if they time their flaps and squeaks correctly, echolocation costs no extra energy. It is an excellent bargain – fly, and get sonar for free.

Other parts of the fossil provide clues about *Onychonycteris*'s lifestyle. Its strong hind legs and wing claws suggest that it was an agile climber and could have scampered through branches on all fours. The proportions of its limbs are closest to the modern mouse-tailed bats, which fly with an undulating style that alternates true flight with gliding. *Onychonycteris* may have also used this efficient tactic and it may have been a stepping stone from gliding to proper continuous flight. But how could *Onychonycteris* have found its way through the air without echolocation to guide it? Perhaps it was a daytime flier and relied on vision to find its way around. Its descendants may have been forced to become nocturnal when the birds came to power towards the end of the dinosaurs' reign, some 65 million years ago. We would normally turn to *Onychonycteris*'s eyes for clues because nocturnal animals that fly by sight, like the flying foxes, have large eye sockets. Unfortunately, that will have to wait until the discovery of new fossils. The top half of the skulls that Simmons used were found crushed. While their ear cavities have already told us much about the evolution of bats, their broken eyes sockets will tell no tales.

The automatic evolution machine

Our bodies are serviced by a huge workforce of enzymes, which speed up the chemical reactions that rage within our cells. These enzymes have been crafted into a vast array of shapes and functions over millions of years of evolution but new ones can be generated on a microchip using the same principles. Early attempts to design proteins from scratch resulted in fairly crude enzymes that were outperformed by nature's more elegant and finely-tuned efforts. Scientists have since developed more efficient artificial enzymes using the same evolutionary principles that generated naturally occurring ones. The process involves mutating an initial pool of enzymes to get a varied mix, testing them for an ability-of-choice, and weeding out the most successful ones for further development. This is laborious and time-consuming work, and it requires the attention of experimenters at every step of the way. It is also not quite as elegant as a biologist might like. If designing proteins from scratch seems a bit like the work of a creator, then checking and steering the development of evolving proteins is rather like the work of an intelligent designer. A more Darwinian system would apply a set of rules to some starting ingredients and let events unfold without tampering. That is exactly what Brian Paegel and Gerald Joyce from the Scripps Research Institute have done. They have created an automatic "evolution machine" that drives the evolution of enzymes without any human direction.[27] All the experimenter has to do is to set some initial conditions, provide the ingredients and switch the machine on.

In their first successful trial, the duo used the machine to develop a more efficient version of an enzyme called an RNA ligase. Most enzymes are proteins made of amino acids, but RNA ligase is just a folded molecule of RNA. Its job is to speed up a chemical reaction which attaches itself to another piece of RNA called a promoter. With a promoter attached to it, the RNA ligase is recognised by two other enzymes that create another copy of it (reverse transcriptase and RNA polymerase). This means that RNA ligases that are good at their job – sticking promoters to themselves – become more abundant and eventually dominate the mix of molecules. The more efficient they are, the more plentiful they become. In a similar way, living things that outcompete their rivals reproduce more quickly and become more numerous in a population.

The evolution machine uses a technique called microfluidics to carefully control the flow of liquids along tiny channels carved onto a microchip. The channels include a main loop, where the ingredients are allowed to mix, and a series of valves that pump liquid in and out of the loop. The whole system is controlled by a computer so Paegel and Joyce could turn their backs on it once it had started. They started with a single RNA ligase that was randomly mutated to give over two billion varieties, all slightly different and some more efficient than the others. This variety provided the raw material for natural selection to work with. This melange of ligases was added to the mixing loop together with promoters, the two other enzymes and a chemical called thiazole orange. When this comes into contact with RNA, it glows with a distinctive orange colour under laser-light, so the more orange the liquids were, the more RNA they contained. As the successful RNA ligases are amplified, they emit a stronger glow and when this reaches 10 times the original level, the machine withdraws 10% of the liquid and keeps it aside. The loop's contents are then flushed out, fresh ingredients are added, the isolated fraction is pumped back in and the cycle begins anew.

At certain points during the experiment, the concentrations of promoter were reduced to keep the evolving populations under pressure; only the most efficient enzymes, which could stick to promoters when they are scarce, would have been amplified. By the end of the experiment, they had to work with about 20 times less material than at the start. Every time the levels of promoter were dropped, the remaining ligases were also removed and mutated again to increase their variety. The experiment carried on for 500 cycles, which together took just under three days to complete. In the end, it produced ligases that were 90 times more efficient than their predecessors. The success of this new breed was the result of 11 mutations that all of them had developed. By individually adding these mutations to the original ligases, and taking them away from the final ones, Paegel and Joyce could test the effects of each change. They found that three of the 11 were particularly important and by themselves, they boosted the enzyme's efficiency by 24 times. Two other mutations were only beneficial in the context of the others and alone, they actually made the enzyme less efficient. These complex traits only turned up at the end of the experiment, during the last few hundred cycles.

It is often (and wrongly) understood that evolution is a wholly random process, which some people use as an excuse to ignore the whole science altogether. But the evolution machine provides a beautiful demonstration of the ways in which evolution is both random and non-random. Paegel and Joyce programmed the machine to select for enzymes that could carry out a certain chemical reaction in the face of falling levels of ingredients. They effectively set the selection pressures that guided the development of the enzymes – that much is not random. But the ways in which the enzymes met the challenge *were* random. Paegel and Joyce claim that the combination of 11 mutations they developed could not have been predicted from the outset. If they ran the experiment again, a wholly different set of mutations might emerge. So far, the machine's use is limited by the very simple nature of the RNA ligase enzyme. We're a long way from using it to create new proteins based on amino acids with more complex functions. Even so, it provides a 'proof-of-principle' that such a machine is possible. Paegel and Joyce claim that the microfluidic systems are so precise and easy to use that they could eventually be used to govern the evolution of proteins, viruses and bacteria. A multitude of experiments could be performed on the face of a single tiny microchip.

Immune snakes outrun toxic newts

The story of evolution is filled with antagonists, be they predators and prey, hosts and parasites, or males and females. These conflicts of interest provide the fuel for 'evolutionary arms races' – cycles of adaptation and counter-adaptation where any advantage gained by one side is rapidly neutralised by a counter-measure from the other. As the Red Queen of Lewis Carroll's Through the Looking Glass said to Alice, "It takes all the running you can do, to keep in the same place." The Red Queen analogy paints a picture of natural foes, wielding perfectly balanced armaments and caught in a perpetual stalemate. But this is an oversimplified view. It is entirely possible for one combatant to develop such a significant advantage that it completely outruns the other and temporarily wins the race. Charles Hanifin from Utah State University has found one such example among garter snakes and newts living along North America's west coast.[28] Even though some of the newts pack one of the most powerful

poisons used by any animal, they still fall prey to garter snakes that have evolved extreme levels of resistance to them. In some locations, the snakes' immunity is so complete that the not a single newt is poisonous enough to overwhelm them.

The three species of newts from the genus *Taricha* defend themselves with a lethal poison called tetrodotoxin. It kills by plugging up molecular pores on the surface of nerve and muscle cells. These pores act as channels for sodium ions and if these ions are denied passage, nerve cells cannot fire and muscles cannot contract. The heart stops, breathing becomes impossible and death soon follows. There is no antidote. The skin of a single newt is laced with enough tetrodotoxin to kill 10-20 humans, or thousands of mice, but not the common garter snake (*Thamnophis sirtalis*; above). Some individuals have become immune to tetrodotoxin, by changing the structure of their sodium channels so that the poison no longer blocks them.

To study the arms race between snake and newt, Hanifin surveyed different populations across their entire shared range, a 2,000 km stretch of land between British Columbia and the southern tip of California. While many arms-race studies look at a single pair of populations, that is a bit like spotlighting on two actors on a crowded stage; instead, Hanifin wanted to watch populations at different stages of escalation across a large geographical area. Together with two Edmund Brodies (Jr and III), he measured the levels of tetrodotoxin in newts from 28 locations across the west coast. They also measured how resistant local snakes were by injecting them with the poison and measuring its effect on their slithering speed. As expected, they found massive differences in both toxicity and resistance. Some populations have not entered the arms race at all; in British Columbia, for example, non-resistant snakes live alongside poisonless newts. In other areas, as the newts become more toxic, the snakes become more resistant and the conflict escalates until both poison and resistance are magnified a thousand-fold.

In general, the most resistant snakes lived alongside the most toxic newts. But Hanifin also found that the animals' abilities were often mismatched and in every single case, it was the snakes that came out ahead. In a third of the locations they sampled, even the least resistant snakes were more than capable of eating the most toxic newts. Taking mouthfuls of one of the most lethal of animal poisons barely slowed them down.

In these locations, the snakes have escaped from the cyclic nature of the evolutionary arms race. Their advantage is so great that there is not a newt toxin they cannot handle, and as such, they are under no impetus to become even more resistant. It seems surprising that the newts never developed an overwhelming advantage themselves. After all, you might assume that they would be under even greater pressure to develop better defences for they stand to lose their lives while the snakes merely risk losing their dinner. But the snakes have a genetic advantage. Their ability to shrug off the effects of tetrodotoxin depends on the structure of their sodium channels and these in turn are governed by a small number of genes. The upshot is that it takes a very small number of simple genetic changes to turn a susceptible snake into a resistant one and these chances can spread rapidly throughout a population.

In at least one group of extremely resistant snakes, the altered sodium channel differs from the basic model by a single amino acid in its entire length. The effect is like making a fort invulnerable by changing the position of a single brick. It is altogether more complicated for newts to evolve more powerful poisons. Some scientists have suggested that the various animals that wield tetrodotoxin may accumulate it from an environmental source rather than making it themselves, and that would limit the amount that an individual could build up. Tetrodotoxin is also so powerful that the newts themselves are not immune to it. They safely store the chemical in their skin but it would be physically impossible for them to house enough to overwhelm the defences of the most resistant snakes.

What happens next is unclear but while the race has been temporarily suspended, it is not over. While the snakes can take a breather from all the relentless innovation, the newts are still very much in the game and under strong pressure to develop even more lethal defences, if they can. Alternatively, the snakes may even find it beneficial to become less resistant. Their altered sodium channels open up an exclusive menu of newts unavailable to other predators, but they carry a cost too. The changes to the snakes' nerves and muscles make them move more slowly and Hanifin speculates that if this drawback is significant enough, the snakes could begin to lose resistance. This de-escalation of arms could bring them back to a level where they could once again be poisoned by the newts, and the race would be rejoined.

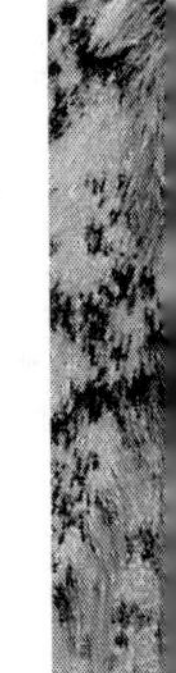

The evolution of languages

Languages are often compared to living species because of the way in which they diverge into new tongues over time in an ever-growing linguistic tree. Their words and grammars change and mutate over time, and new versions slowly rise to dominance while others face extinction. Some critics have claimed that this comparison between linguistic and biological evolution is a superficial one, a nice metaphor but nothing more. But more and more, studies are showing that the evolution of languages mirrors that of living things in many key ways. For example, a group of researchers led by Quentin Atkinson, now at the University of Oxford, found that the birth of new languages is accompanied by a burst of rapid evolution consisting of large changes in vocabulary, followed by long periods of relatively slower change.[29] This stop-and-start pace uncannily echoes a long-standing theory in biology known as 'punctuated equilibrium'. The theory's followers claim that life on Earth also evolved at an uneven pace, full of rapid bursts and slow crawls. Famously championed by the late Stephen Jay Gould, the punctuated equilibrium theory suggests that most species change very little over time and big evolutionary changes are concentrated at rare moments where new species branch off from existing lineages. Together with colleagues from the US and New Zealand, Atkinson found similar patterns in three of the world's largest families of languages.

They compared lists of words from the Indo-European group, which include English and Hindi; the Bantu group, consisting of several hundred African languages; and the Austronesian group, which includes over a thousand tongues from Indonesia, Papua New Guinea and the Polynesian islands. Between them, these three families account for a third of all the world's languages. Of course, languages borrow words from each other all the time and indeed, 50% of English words are loans from French and Latin. That was a potential pitfall of the study and Atkinson avoided it by only considering basic words such as numerals, body parts and pronouns that are very unlikely to have been co-opted from another tongue.

For each group, Atkinson built a family tree showing how newer languages split off from ancestral ones. The trees mirrored those that biologists use to chart the evolutionary relationships between species. In the model, the birth of new languages is represented by new branches on the tree and the length of each branch depends on the difference in vocabulary between the new tongue and its parent one. The greater the changes, the longer the branch. In each family tree, Atkinson saw that the parts of the tree with the most branches also had the longest ones. So groups that spawned the highest number of new languages also diverged most significantly in their repertoire of words. That is the pattern you would expect if the birth of new languages triggered bursts of rapid evolution. If the pace of evolution was more constant, the number of branch points would have no effect on overall branch length.

These rapid bursts accounted for 31% of the vocabulary differences between Bantu speakers, 21% of the differences in Indo-European languages and 10% of the variation in the Austronesian group. For comparison, team estimated that these fast and slow evolutionary cycles explained about 22% of the genetic differences between biological species. As they split from each other, new sister tongues begin to adopt new words at a fast pace and these are probably accompanied by equally quick changes in pronunciation, spelling and grammar. As their identities become clearer, the pace of change slows. Atkinson thinks that this process happens when different groups of people try to establish distinct social identities by exaggerating differences in language. American English may have developed along these lines and the need for a unique identity was at the forefront of Noah Webster's mind when he published his first *American Dictionary of the English Language* in 1828. "As an independent nation, our honour requires us to have a system of our own, in language as well as government," he said.

Languages also have their own fossil record. The real biological record is the ever-growing collection of bones from extinct species that unveil how life on earth changed and adapted over millions of years. The linguistic equivalents are old texts like the Canterbury Tales that preserve the existence of words that used to be commonplace before they lost a linguistic Darwinian conflict with more popular forms. Erez Lieberman, Martin Nowak and colleagues from Harvard University have are looked at this record to mathematically model how our verbs evolved and how they will change in the future.[30]

Today, the majority of English verbs take the suffix '-ed' in their past tense versions. Sitting alongside these regular verbs like 'talked' or 'typed' are irregular ones that obey more antiquated rules (like 'sang/sung' or 'drank/drunk') or obey no rules at all (like 'went' and 'had'). In the Old English of *Beowulf*, seven different rules competed for governance of English verbs, and only about 75% followed the "-ed" rule. As the centuries ticked by, the irregular verbs became fewer and far between. With new additions to the lexicon taking on the standard regular form ('googled' and 'emailed'), the irregulars face massive pressure to regularise and conform.

Today, less than 3% of verbs are irregular but they wield a disproportionate power. The ten most commonly used English verbs – be, have, do, go say, can, will, see, take and get – are all irregular. Lieberman found that this is because irregular verbs are weeded out much more slowly if they are commonly used.

To get by, speakers have to use common verbs correctly. More obscure irregular verbs, however, are less readily learned and more easily forgotten, and their misuse is less frequently corrected. That creates a situation where 'mutant' versions that obey the regular "-ed" rule can creep in and start taking over. Lieberman charted the progress of 177 irregular verbs from the 9th century Old English of *Beowulf*, to the 13th century Middle English of Chaucer's *Canterbury Tales*, to the modern 21st century English of Harry Potter. Today, only 98 of these are still irregular; many formerly irregular verbs such as 'laugh' and 'help' have put on new regular guises. He used the CELEX corpus – a massive online database of modern texts – to work out the frequency of these verbs in modern English.

Frequency	Verbs	Regularization (%)
1 in 10 or more	be, have	0
1 in 100 to 1 in 10	come, do, find, get, give, go, know, say, see, take, think	0
1 in 1,000 to 1 in 100	begin, break, bring, buy, choose, draw, drink, drive, eat, fall, fight, forget, grow, hang, **help**, hold, leave, let, lie, lose, **reach**, rise, run, seek, set, shake, sit, sleep, speak, stand, teach, throw, understand, **walk**, win, **work**, write	10
1 in 10,000 to 1 in 1,000	arise, **bake**, bear, beat, bind, bite, blow, **bow**, burn, burst, **carve**, **chew**, **climb**, cling, creep, **dare**, dig, **drag**, flee, **float**, **flow**, fly, **fold**, freeze, grind, leap, lend, **lock**, **melt**, **reckon**, ride, **rush**, **shape**, shine, shoot, shrink, **sigh**, sing, sink, slide, **slip**, **smoke**, spin, spring, **starve**, steal, **step**, **stretch**, strike, **stroke**, **suck**, **swallow**, swear, sweep, swim, swing, tear, wake, **wash**, weave, weep, **weigh**, wind, **yell**, **yield**	43
1 in 100,000 to 1 in 10,000	**bark**, **bellow**, bid, **blend**, **braid**, **brew**, **cleave**, **cringe**, **crow**, dive, **drip**, **fare**, **fret**, **glide**, **gnaw**, **grip**, heave, **knead**, **low**, **milk**, **mourn**, **mow**, **prescribe**, **redden**, **reek**, **row**, **scrape**, **seethe**, shear, shed, **shove**, slay, slit, **smite**, sow, **span**, **spurn**, sting, stink, strew, stride, swell, **tread**, **uproot**, **wade**, **warp**, **wax**, **wield**, wring, **writhe**	72
1 in 1,000,000 to 1 in 100,000	**bide**, **chide**, **delve**, **flay**, **hew**, **rue**, **shrive**, slink, **snip**, **spew**, **sup**, **wreak**	91

Table 1. Some irregular English verbs. Bold words were once irregular but have since become regularised

Amazingly, he found that this frequency affects the way that irregular verbs disappear according to a very simple and mathematical formula. They regularise in a way that is "inversely proportional to the square root of their frequency". This means that if they are used 100 times less frequently, they will regularise 10 times as fast and if they are used 10,000 times less frequently, they will regularise 100 times as fast. As Lieberman says, "We measured something no one really thought could be measured, and got a striking and beautiful result." Using this model, the team managed to estimate how much staying power the remaining irregular verbs have and assigned them 'half-lives' just as they would to radioactive isotopes that decay over time.

The two most common irregulars – 'be' and 'have' – crop up once or more in every ten words and have half-lives of over 38,000 years. That is such a long time that they are effectively immune to regularity and are unlikely to change. Less common verbs like 'dive' and 'tread' only turn up once in every 10,000-100,000 words. They have much shorter half-lives of 700 years and for them, regularisation is a more imminent prospect. Out of the 98 remaining irregular verbs examined in the study, a further 16 will probably have adopted the '-ed' ending by 2500.

Which will be next? Lieberman has his speculative sights set on 'wed'. It is one of the least commonly used of modern irregular verbs and the past form 'wed' will soon be replaced with 'wedded'. As he jokes, "Now is your last chance to be a 'newly wed'. The married couples of the future can only hope for 'wedded' bliss". That little jibe highlights the greatest strength of this paper – not the striking and elegant results but its delightful turns of phrase. Suitably for a study about language, he describes his results in pithy and measured language. Observe, for example, his concluding paragraph:

> "In previous millennia, many rules vied for control of English language conjugation and fossils of those rules remain to this day. Yet, from this primordial soup of conjugations, the suffix '-ed' emerged triumphant. The competing rules are long dead, and unfamiliar even to well-educated native speakers. These rules disappeared because of the gradual erosion of their instances by a process that we call regularisation. But regularity is not the default state of a language – a rule is the tombstone of a thousand exceptions."

Regularised verbs are not the only structures to crop up as languages evolve – the changing dialects also have a tendency to become more structured and easier to learn. Regardless of school lessons and textbooks, most of the features of the languages we speak are learned by listening to the words of native speakers. Their sentences convey their thoughts, but they also hint at the structure of the language they are spoken in. That allows people who are learning a new language to infer something about its structure by listening to the way its sentences are put together. In the past, computer models have shown that behaviours like these, which are passed on through repeated cycles of observation and learning, eventually evolve to become easier to learn. But these simplified models are fairly removed from realistic learning. In his book on language evolution, Derek Bickerton described these models as a "case of looking for your car-keys where the street-lamps are." The big question is whether real languages have also evolved in this way?

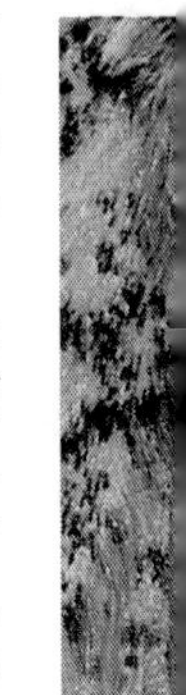

We really need experiments that test the adaptations that languages pick up over time, using the brains of living people rather than the software of computers. And that is exactly what Simon Kirby and Hannah Cornish have done. Together with Kenny Smith at Northumbria University, they have been watching evolution occur within the confines of their laboratory, but in languages rather than bodies or genes. In doing so, they have provided the first experimental evidence that as languages are passed on, they evolve structures that make them easier to transmit effectively. The team tracked the progress of artificial languages as they passed down a chain of volunteers.[31] They found that in just ten iterations, the made-up tongues became more structured and easier to learn. What is more, these adaptive features arose without specific plans or designs on the part of the speakers. The appearance of design without the guiding will of any designer is another trait that offers compelling parallels to biological evolution.

The trio asked a group of 80 volunteers to learn an "alien language" that described different visuals. The visuals were 27 combinations of shapes (square, circle or triangle) that had a specific colour (red, black or blue) and moved in a specific way (horizontal, bouncing or spiralling). Each of these was attached to a randomly made-up word of 2-4 syllables. After training, the volunteers were shown all the images and asked to say what word the aliens would give to describe each one, with the catch that they had only been trained on about half of the pairs. The first volunteer's responses were used to train and test the second; their answers were used on the third and so on. This was done for four separate chains using different starting words. In each one, the languages had clearly become easier to learn by the 10th round, with the later learners making fewer and fewer mistakes than those before them. A couple of the languages were even transmitted perfectly by the end; people came up with exactly the same set of 27 words as their predecessors, *even though they had not seen half of the word-image pairs before.*

The languages also became much more structured over time. At the start, every image had its own unique word that could only be learned by rote, as nothing in the words themselves gave any clue to the meanings they conveyed. But by the end, single words were reused to express more than one meaning and the total number of words had plummeted to a mere handful from the original count of 27. At first glance, you might think that this shedding of words explains why the later participants made fewer mistake. But if the languages simplified in a random way, the volunteers still shouldn't have been able to achieve perfect scores. For example, if the same word means a black, bouncing circle and a red, spiralling square, you could not draw any general conclusions about its use, and you'd be back to learning by rote again. Clearly, that was not possible for the volunteers, who were only trained on half of the word-image pairs. The key to the perfect scores was the fact that the languages became simpler in *systematic ways.* For example, by the 4th round in one of the chains, the word *tuge* came to mean any horizontally moving object, regardless of shape or colour. By the 6th round, *poi* meant any bouncing object and at the end, the spiralling objects had three different words depending on their shape. It was precisely this structure that allowed people could deduce the words for objects they had never seen before with exacting precision.

A nice first result, but the ambiguity of the end products is a problem. Languages are tools for communication and no matter how easy one is to learn, it would be useless if it could not put across different ideas clearly. In short, languages need to be *expressive.* But with one small tweak to their experiment, Kirby and Cornish showed that languages can do this while still evolving to be learnable and structured. This time round, they filtered the pairs between rounds so that if multiple images were represented by the same word, all but one of them was removed. This meant that at every stage, each word could only have a single 'meaning' and with this change, the languages remained relatively wordy till the end.

However, they still became easier to learn, with volunteers making fewer mistakes later on (although there were no perfect scores). And they *still* became increasingly structured but in a very different way. This time, the structure lay *in the way the words were put together.* By the sixth round in every single chain, each word was made up of three parts that referred to the object's shape, colour and movement. For example, in one chain, words that described blue shapes started with *n*, those for black shapes started with *l* and those for red shapes began with *r.* Likewise, words that ended in *ki* described horizontal movement, those ending in *plo* were for bouncing shapes and those ending in *pilu* meant spiralling. So a red, bouncing circle might be *rehoplo*, while a blue, spiralling square would be *nepilu.*

Kirby and Cornish claim that the results of both experiments are relevant to the way that real languages work. In their first test, the artificial dialects became more structured as words took on a methodical collection of meanings. The same thing happens in real life. For example, the majority of nouns refer to multiple objects, while only proper nouns refer to individual things. All cats, for example, are known as cats, regardless of their shape, colour or movement. In the second experiment, words became structured by ascribing specific meaning to components that can then be combined. Natural languages are rife with these hierarchies, at both the levels of words and sentences. In the experiment, they emerged spontaneously, but only because Kirby and Cornish deliberately removed multiple meanings between each round. That is certainly an artificial step, but they argue that it reflects the pressures that real-world languages face – to be both expressive and easy to transmit.

In both cases, it was vitally important that from the volunteers' perspective, they were doing the same task. They were not told that they were working from the outputs of other people who had gone before them, nor did they guess what was going on. They were not trying to improve on the language in any way; their goal was to reproduce it as accurately as they could and many of them did not even being realise that they were tested on images that they had not seen during the training. And that brings us to the final similarity between the world of words and species – in both cases, evolution can lead to the appearance of design without a designer.

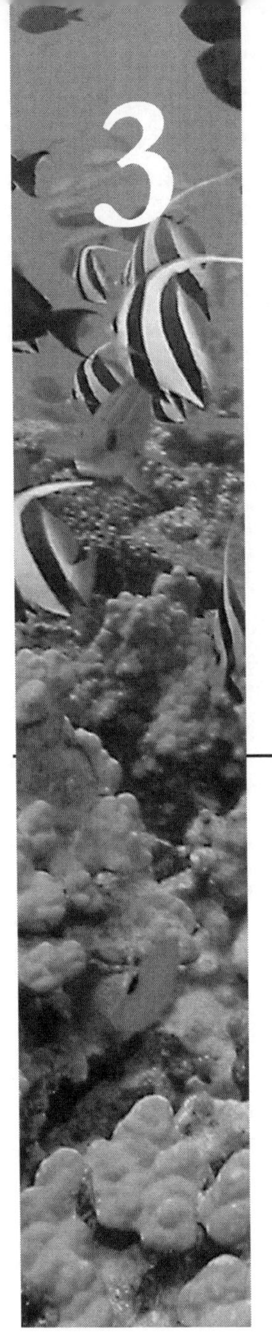

Living ripples

Climate-changing beetles,
why castrators succeed and
why chillies are hot

Why chillies are hot

Food from countries all over the world owes a lot of its flavour to a fungus. The species in question is not one of the many edible mushrooms used for cooking, the baker's yeast that gives rise to bread, or the moulds that spread through blue cheese. It is a little known species called *Fusarium semitectum* and its role has only just been discovered – it is the driving force behind the fiery heat of chillies. Chillies were one of the first plants from the New World to be domesticated and they have spread all over the world since. Just today, a quarter of the world's population has eaten something flavoured with chillies, be it salsa, curries or stir-fries. The chemicals behind their eye-watering bite are called capsaicinoids, and they are unique to this group of plants. Why did they evolve?

No matter what cookbooks and supermarkets tell you, chillies are fruits – fleshy parts of plants that contain seeds. Like all fruits, they serve to reward pollinators for spreading a plant's seeds, using the currency of nutritious pulp. But these same nutrients make fruits a ripe target for other unwelcome species that are not involved in pollination or, even worse, eat the seeds. To deter these unwanted fruit-eaters, plants defend themselves with toxic chemicals, such as capsaicinoids. Take the chillies – their seeds pass through the guts of birds unharmed but they fail to sprout after being eaten by mammals. Birds, however, lack the receptors that sense capsaicinoids and trigger the burning sensations that mammals like ourselves experience all too readily. So it is tempting to view capsaicinoids as a mammal-deterrent, but Joshua Tewksbury at the University of Washington found that this is not the whole story.

Working in Bolivia, Tewksbury found that while mammals are an occasional nuisance, the biggest threat that chillies face are microbes, which are also after the sugars and nutrients they contain.[32] One species in particular – *Fusarium semitectum* – rots the fruits and kills the seeds, and it is the only major cause of damage to the plant. If the capsaicinoids are intended to defend the plant against a threat, then surely it was this fungus. To test this idea, the research team compared chilli plants across 1,000 square miles of Bolivian forest. At the south-western end of this area, almost every single chilli is loaded with capsaicinoids and packs jalapeno-level heat. Head up north and east and the fruits become less and less pungent so that at the north-eastern corner, more than 70% of chillies lack any capsaicinoids at all. Those that still have them pack less than a third of the levels found in their spicier peers.

Tewksbury compared the ability of these different fruits to resist infection by placing cages over them that would block out the advances of birds but let in small bugs like aphids and treehoppers. Their biting mouthparts are the vessels that *Fusarium* stows away on. When they pierce the chillies to suck their juices, they also seed them with fungus and their bites leave a telltale scar on the fruit's surface. These experiments revealed that the north-eastern chillies weren't just weaker in taste – they were also weaker against *Fusarium.* Lab experiments have revealed that capsaicinoids kill off a variety of microbes, and compared to the weak capsaicinoid-free chillies, those that were loaded with these chemicals had half the amount of seeds that were infected by the fungus. To rule out the possibility that some other factor was responsible for protecting against *Fusarium*, the researchers mixed up different pots of jam that mimicked the nutrients of chillies but had differing levels of capsaicinoids. As expected, they found that the fungus was least able to grow in pots with the highest concentrations of these pungent chemicals. In fact, at the concentrations of capsaicinoids found in the spiciest natural peppers, the fungus's growth was slashed exactly by half, showing that these chemicals and they alone were responsible for safeguarding the chillies.

If capsaicinoids are so important, why are some of Bolivia's chillies weak? The answer is simple – *Fusarium,* and the bugs that spread it, are rare in these areas and the fruits have far less need for defensive measures. In the northeast, where chillies are milder, Tewksbury found very few of the bug-made scars that indicate entry by the fungus, whereas the hotter chillies in the southwest were rife with them. So the presence of fungi and bugs causes some chillies to defend themselves with chemical weapons, and this evolutionary pressure is at least partially responsible for the wide variations in chilli strength. It might also explain why humans started eating chillies in the first place. Tewksbury suggests that our ancestors may have first started using and domesticating chillies for their ability to kill of nasty microbes. Before refrigerators became commonplace, the fungi and bacteria lurking within our food were a massive problem, and it is possible that our ancestors developed a taste for spicy food precisely because the chemicals that make our eyes water also cause trouble for contaminating bugs.

A perfect storm for epidemics

In 1994, a third of the lions in the Serengeti were killed off by a virus. They succumbed to a massive epidemic of canine distemper virus (CDV), an often fatal infection that affects a wide range of meat-eating mammals. Seven years later, a similar epidemic slashed the lion population in the nearby Ngorongoro Crater in Tanzania and again, CDV was undoubtedly involved. But the scale of the deaths in both instances was completely unprecedented. What was it about the two epidemics that claimed so many lion lives? A team of scientific detectives led by Linda Munson from the University of California, Davis have solved the mystery. They found that no single perpetrator was responsible for the lions' deaths. Instead, there were multiple culprits – a combination of viral infections, blood parasites and extreme climate changes.[33] Heavy droughts had triggered a complex chain of events that left the lions riddled with an unusually high burden of parasites, a problem compounded by CDV infections, which depleted their immune system and left them vulnerable. It was a "perfect storm" of events that sealed their fate.

Munson's international team analysed blood samples that were taken from 510 lions between 1984 and 2007. They found that during this period, seven CDV epidemics had actually swept through the lion population, some of which infected 95% of individuals. However, five of these were silent and had little effect on the lions' death rates. Munson's team discovered that the blood samples taken from lions during the two CDV epidemics had extraordinarily high levels of a blood parasite called *Babesia.* They found the same thing in samples collected just after the 2001 epidemic from 10 anaesthetised lions and two fresh carcasses. These lions were carrying some of the highest levels of *Babesia* ever measured, anywhere from 34-60 times the usual tally, and the two carcasses showed signs of babesiosis, the malaria-like disease that *Babesia* causes. The two infections – *Babesia* and CDV – were clearly interacting in some way. Lion prides that were infected with higher levels of CDV also had higher levels of *Babesia* and prides that

were most afflicted by *Babesia* had the highest death rates. That makes sense, for CDV suppresses the immune system and would make lions more vulnerable to parasites. Indeed, the only pride that remained completely CDV-free in 1994 were not strongly affected by the parasite and did not show any spike in deaths.

But even this partnership cannot explain the scale of death in the two epidemics. During the silent epidemics, the lions' immune systems would also have taken a hit and yet, *Babesia* failed to run riot in their blood. Munson found that the missing piece of the puzzle was the environment. The common factor that linked the two catastrophic epidemics was severe drought in the year before, followed by heavy rains. The 1993 drought in the Serengeti was the worst in over 40 years and the drought that hit the Ngorongoro Crater in 2000 was equally extreme. Droughts are good news for the blood-sucking ticks that spread *Babesia*, because the plant-eaters they feed on are weaker. It is less good for the plant-eaters, and in 2001, several black rhinos died in the Crater due parasites transmitted by massive tick infestations. Cape buffalo also suffered and they are a high on a lion's menu. When the rains returned after both droughts, the buffalo became riddled with ticks and died off in large numbers and that provided the lions with unprecedented helpings of tick-infested flesh. Munson's team found that lions had much higher counts of *Babesia* if they fed off buffalo during the epidemics, and especially if they were also infected with CDV. Two prides that contracted CDV but did not feed on buffalo in 1994, survived the epidemic with few parasites and few deaths.

The study is a beautiful example of how an epidemic can be triggered by a multiple-whammy of different factors, including environmental changes and side-by-side infections. Drought weakened the herbivores, which left them vulnerable to ticks, which exposed lions to blood parasites, which ran rampant among bodies whose immune systems had already been suppressed by CDV.

Even more factors may have contributed to the lions' downfall, especially in the Ngorongoro Crater. The animals there live in dense groups, ideal for the transmission of parasites. The Crater lions are very inbred and may be more susceptible to infections, parasitic flies that cause skin ulcers in lions also flourished in the post-drought downpours, and human interventions in grass fires may have contributed to the rise in tick numbers in 2001 (a policy of controlled burning has since been put in place).

Similar webs of events may be happening elsewhere in the world. Since 2006, honeybees across Europe and North America have been mysteriously disappearing and last year, a breakthrough paper showed that this "colony collapse disorder" was caused by a virus but one that is unlikely to act alone. The scientists who made the discovery suggested that a blood-sucking mite called *Varroa* could make honeybees more susceptible to infection by weakening their immune systems. Even the chemicals used to control the mites, as well as pesticides sprayed on pollinated plants, could add to the bee's health burden.

The big worry is that perfect storms like these will become ever more common as global warming leads to large shifts in climate. These changes have the potential to worsen the effects of existing diseases, expand their range or cause stable populations to succumb to combinations of disease that they cannot cope with.

Castrating the way to success

To a science-fiction filmmaker, the concept of being controlled by unseen forces is creative gold, but for the rest of us, it is a fairly unsettling prospect. But like it or not, it is clear that parasites – creatures that live off (and often control) the bodies of others – are an integral part of the world we live in and carry an influence that far exceeds their small size. Now, a painstaking survey of three river estuaries shows that parasites do indeed punch above their weight, and they are not slouches in that department either. Despite their tiny size, their combined mass eclipsed that of the top predators in the area and their influence extended wider still. It is a parasite's world and we're just living in it.

Over five years, Armand Kuris and Ryan Hechinger from the University of California, Santa Barbara led an exhaustive census of life in three Californian estuaries.[34] At 69 different sites, they assessed almost 200 species of free-living animals, from high-flying birds to burrowing shrimps, as well as the 138 species of parasites hitching a lifestyle on their bodies. Using a wonderfully old-school combo of nets, binoculars, meshes, sieves and scales, the team collected samples of the estuaries' inhabitants and dissected over 17,000 individual animals. Their bodies were dissected and their parasites removed, classified and weighed. Snails, crabs and bivalves (hinged shellfish) made up the bulk of the local fauna but every single animal group living in the estuaries was infected with parasites of some description. On average, these freeloaders made up about 2% of the weight of their hosts. That may not seem like much, but it is about ten times more than previous estimates. It was always assumed that parasites, being very small, would weigh next-to-nothing despite their weighty influence; now, we know that is not true.

Tapeworms and parasitic crustaceans were very common, but lording over all of them were the trematodes or flukes. This group of parasitic worms alone weighed as much as the local fish and outweighed the birds – the local top predators – by three to nine times. Pound for pound, the estuaries were home to more parasite tissue than bird tissue. Kuris and Hechinger also found that some parasites use particularly successful strategies and the dominant group were those that castrate their hosts by infecting and shutting down their reproductive systems. Castrators are the ultimate body-snatchers for their neutered hosts have zero chance of passing their genes down to the next generation. In effect, their usurped bodies become living shells for the parasites' own genes. It is a superb example of what Richard Dawkins calls "extended phenotypes" – influences than an animal's genes have beyond the body that they inhabit, like the dams of beavers or the cities of humans.

Of the many strategies used by local parasites, castration was by far the most successful, so much so that the combined weight of castrated bivalves, snails, shrimps and crabs was almost the equal of their fertile peers. Remember that these were the most common animal groups around, and take a second to reflect on that sobering statistic. It means that a massive proportion of the flesh in the estuaries is under parasite control and otherwise useless to their owners. Take the California horn snail, *Cerithidea californica.* It is the most abundant single species in the area that Kuris and Hechinger studied and it is infected by 18 species of trematode. Almost all of the largest snails were castrated and on average, a massive 22% of an infected snail's body was made up of parasites. The snails had been converted into trematode factories. Every day, the infected snails released a swarm of cercariae – a part of the trematode's life cycle that swims with a tadpole-like tail. Each of these lives for just a day, but the snails discharge so many into the water that the combined weight of the cercariae alone also dwarfs the mass of local birds by three to ten times.

The snail's tale illustrates that parasites have influence far beyond their already considerable weight. They may have the substantial combined mass of larger animals, but because they do not need to move or forage, they also enjoy the high productivity typically enjoyed by microscopic organisms. Kuris and Hechinger admit that they only looked at one specific ecosystem, but they believe that their estimates will apply elsewhere and that they are, if anything, too conservative. If they are right, no consideration of food webs or ecology would be complete without taking the power of the local parasites into account.

Beetlemania

In the story of climate change, humans and the carbon dioxide we pump into the atmosphere are the villains of the piece. Now, it seems that we have an accomplice and a most unexpected one at that. It lives in the pine forests of North America and even though it measures just five millimetres in length, it is turning these woods from carbon sinks into carbon sources. It is the mountain pine beetle.

The beetle (*Dendroctonus ponderosae*) bores into pine trees and feeds from nutrient-carrying vessels called phloem. It also lays its eggs there. Once a beetle has colonised a pine, it pumps out pheromones that attract others, which descend on the tree en masse and overwhelm its defences. The infestation damages the phloem to such an extent that the tree effectively dies of malnutrition within weeks. A single beetle can live for up to a year, giving it plenty of time to damage several trees. Normally, the beetles target weak, old or dying trees and in this capacity, they accelerate the growth of young trees. But occasionally, their numbers erupt in large-scale outbreaks and British Columbia is currently facing the largest one ever recorded. They have infected an area of forest the size of Greece and the scale of the epidemic is ten times worse than any previous incident. The damage is even obvious from the air, for the needles of infected trees turn red within the first year and gray as they succumb further.

The outbreak is obviously bad news for the local ecosystem and the forestry industry alike but Werner Kurz from the Canadian Forest Service has found that the beetle's actions has a subtler negative impact – it will transform the local forest from a carbon sink (which takes in more carbon dioxide than it gives out) into a carbon source (which does the opposite). Dead trees are in no position to soak up carbon dioxide from the air, and their decay will release even more carbon back into their environment.

Kurz modelled the effects of the beetles in the south-central region of British Columbia between 2000 and 2020.[35] He took a wide range of factors into consideration including growth, leaf litter, decay, forest fires and deforestation, and ran 100 different simulations under a range of different conditions. For the first two years of the simulation, the forest was a carbon sink, soaking in half a million tonnes of carbon every year. After 2002, things changed and the falling forests turned into sources of carbon, giving off an average of 17.6 million tonnes every year between 2003 and 2020. The beetles' tree-killing antics will result in the release of about 270 million tonnes of carbon over the 21 year period covered in the simulation. The forestry industry is making things worse by managing the infestation with "salvage-logging", felling infected trees to prevent the spread of the beetles and to recover some economic value from the timber. This harvest will add another 50 million tonnes of carbon to the beetles' direct tally, although only a small proportion of this will be released into the atmosphere.

The effects of these tiny insects match those of more usual suspects. Kurz predicts that this single outbreak will release as much carbon dioxide over 21 years as Canada's entire transportation sector does in five. And the annual amount of carbon released by the beetles in this small patch of British Columbia is similar to the amount given off by forest fires over *all* of Canada during earlier decades. However, the beetles hardly absolve Canada of responsibility over our changing climate, for it was probably climate change that triggered their extraordinary outbreak in the first place. Cold snaps kill the majority of beetles. But over the past few decades, global warming has shifted British Columbia's climate toward milder winters and hotter, drier summers – conditions that the beetles have been quick to exploit by expanding their range, and colonising forests further north, and at higher altitudes.

The problem is not restricted to British Columbia. The mountain pine beetle is wreaking havoc in other parts of North America, and elsewhere, warmer climes have allowed spruce beetles to reproduce in half their normal time, with severe consequences for spruce forests. Kurz's results suggest that infestations of tree-boring insects all over the world could be driving vicious cycles of larger infestations, increased carbon emissions and rising temperatures. There is one positive note. Kurz's simulations suggest that we are currently experiencing the peak of the epidemic. Using data on the beetle populations, the number of remaining host trees and the judgments of entomologists, Kurz predicts that from 2009, the beetles will start literally eating themselves out of house and home. They will have damaged such a large area of pine forest that their own numbers will start to fall. Even so, the impact of the mountain pine beetles will last much longer than their population. By 2020, the forest's productivity will have started to recover, but it still will not have returned to pre-outbreak conditions. Kurz feels that the forests *will* eventually recover but he feels uneasy about making projections beyond the 2020 mark, especially since the area is predicted to go through even larger climate upheavals in the future.

Doom or hope for coral reefs?

If you have never had the pleasure of swimming among a coral reef, you might want to get your chance sooner rather than later. This year, scientists published the first comprehensive global assessment of the status of the world's reef-building corals, and its results do not make for comforting reading. Almost a third of the 700-plus species surveyed face extinction; no group of land-living species, except possibly for the amphibians, are this threatened. A team of 39 scientists led by Ken Carpenter, director of the Global Marine Species Assessment gauged the extinction risk faced by the world's corals by using the International Union for Conservation of Nature's (IUCN) famous Red List Criteria.[36] These criteria measure extinction risk by looking at how quickly a population's size falls over time. Those sorts of rigorous census data simply are not available for most corals, so Carpenter's team settled for the next-best alternative – the rate at which a species' habitat is lost within its known range. The results were adjusted for traits such as the life cycle of each species and how resilient they are to habitat loss. The results showed that the outlook for corals has worsened considerably in just the last 10 years. The team looked at the fates of 704 species and deemed that 176 were Near-Threatened, 201 were Vulnerable, 25 were Endangered and five poor species were Critically Endangered. Using earlier data, the team found that had the analysis been done in 1998 (before a mass "bleaching" event killed off large swathes of coral), only 20 species would have been classified as Near-Threatened and only 13 would have made it into the more severe categories.

It goes without saying that the extinction of a third of the world's reef-building corals would be nothing short of an ecological disaster. Coral reefs harbour the greatest diversity of life of any aquatic ecosystem. Entire assemblages of species depend on their intricate structures for shelter and food. If the corals die, so too will the creatures that depend on them and at last count, about a quarter of oceanic species live in coral reefs. Humans will suffer too. Aside from the loss of some of the planet's most visually striking habitats, coastal economies will suffer as falling biodiversity feeds fewer mouths and dead, colourless reefs attract fewer visitors. Reefs also act as physical barriers that protect coastal communities from the threats of erosion, floods and storms. It is ironic then that most of the threats that corals face are man-made, and they come in forms both global and local.

Close to home, our impact has been more direct. Destructive fishing methods like trawling wreck them directly, while increased sedimentation and pollution lower the quality of the surrounding water and render them vulnerable to disease. On a worldwide scale, the threats faced by corals are just as great. As rising carbon dioxide levels warm the globe, the oceans become both hotter and more acidic. Corals are particularly vulnerable to climate change, for unlike most other animals, they do not have the option of getting up and moving to areas with more hospitable temperatures. The heat causes many species to expel the algae that normally live inside their limestone shells. The algae are symbionts, providing the corals with energy from photosynthesis; they also give the corals their resplendent colours. Without these lodgers, the corals' hues fade (known as "bleaching") along with their energy supply. Meanwhile, the increasingly acidic water depletes valuable carbonate ions that corals need to build their mighty calcium carbonate (limestone) reefs. Like builders with no mortar, they find it hard to erect their homes. Both these dangers – bleaching and acidification – make corals more vulnerable to other threats including direct human damage, disease and the ravenous coral-munching crown-of-thorns starfish (more on them later). The dangers facing corals have a nasty tendency to magnify each others' impacts. This is especially apparent in the triangle of water bordered by Malaysia, Indonesia and the Philippines, where dense human populations exacerbate the effects of changing climate and where the greatest proportion of vulnerable species live.

Not all corals are equal in their vulnerability. Three groups – the Acroporidae, Dendrophyllidae and Euphyllidae – are at greatest peril, due in part to their susceptibility to bleaching. About half of all species have been classified as Threatened or worse and Caribbean reefs have been severely hit by the dramatic decline of two of these – the iconic staghorn coral (*Acropora cervicornis*; above left) and the elkhorn coral (*Acropora palmata*; above right). The corals are an ancient group that have persisted through tens of millions of years of shifting climate. No doubt climate change deniers will point to their dogged perseverance as cause for relaxation. And yet, there is plenty of evidence that several coral species *have* gone extinct in the past – indeed, 45% of all corals died off at the same time that the dinosaurs waved their last goodbyes, and those that relied on symbiotic algae were the hardest hit. Reefs are built over millions of years and it may

take a very long time indeed for corals to recover from the damage that has already been visited upon them. One Caribbean species – the boulder star coral (*Montastraea annularis*) illustrates this point. It is the largest and most abundant reef-building coral in the region but in many reefs, it has relinquished its dominance. It grows slowly but dies easily; within mere months, bouts of disease can kill off colonies that took 500 years to grow. Reefs in the Pacific and Indian Oceans are showing signs of recovery from recent warming events but even though 6% of reefs recovered from the 1998 crisis, a larger 16% were irreversibly destroyed. Their continued survival depends on how often similar events will strike in the future. If bleaching events become more and more common, their decline may become irreversible. Carpenter's paper ends with the following sobering sentence:

> "Whether corals actually go extinct this century will depend on the continued severity of climate change, extent of other environmental disturbances"

There are signs of hope though. In 2004, the Australian Government turned a third of the Great Barrier Reef into the largest network of no-fishing zones in the world. All fishing was banned in an area of sea just smaller than England. It was a bold and controversial political move – jobs and livelihoods, it was said, were on the line. But the plan went ahead and in just a few years, there are signs that it is working. One of the reef's most heavily fished species – the coral trout- is enjoying a dramatic comeback, thanks to this most ambitious of marine conservation projects.

The Great Barrier Reef is the largest living colony of animals on the planet and home to thousands of species. It has been part of a marine park since the 1970s but there were concerns that it was not adequately protected. After all, most of the area was still open to destructive trawling vessels. Due to mounting concerns, the Government launched a massive consultation involving scientists, policy-makers and local communities. The result was to turn 33% of the park into a network of no-take marine reserves, a massive increase from the tiny 5% that had previously been protected. The goal was simple – to prevent the continuing degradation of one of the world's most important marine ecosystems, and to encourage a diverse range of species to return to habitats in decline. The consultation also identified 70 unique habitats or "bioregions" within the Reef's waters, each with its own unique physical and environmental traits and its own cadre of species. From reefs to grass beds to sponge gardens, at least a fifth of the area within each of these zones was protected.

After just a few years, the initiative is already starting to work. Garry Russ and colleagues from James Cook University and the Australian Institute of Marine Science found that one fish in particular – the coral trout – has enjoyed a tremendous comeback since the reserves were set up.[37] The coral trout is a visually striking fish, whose red body is covered in blue spots. Unfortunately, it's also tasty and commands good prices both locally and internationally, which has made it one of the major targets for all sectors of the fishing industry. Before the no-take reserves were set up, Russ's team did a census of marine life in 46 sites that would fall within their boundaries and 46 sites that would lie beyond them. About 18-24 months after the fishing bans were put in place, they repeated their counts and found that the cod numbers had significantly increased. Near the Palm and Whitsunday Islands and in four open offshore reefs, their numbers shot up by about 55-65% in the no-take reserves, but stayed the same in areas open to fishing. The team also found that coral trout numbers did not decline in areas outside the protected zones, which refutes claims made at the time that the no-take reserves would just displace overfishing to other parts of the reef. When the reserves were introduced, the Government also brought in a catch quota to prevent exactly that from happening. This measure too, it seems, has worked. The consistent nature of the coral trout's population rise is extremely encouraging, especially since it's happening over 1,700 km of ocean.

For marine conservation, a positive response like this at such a large scale is completely unheard of. It is the best evidence yet that no-take marine reserves are a good idea. To date, a mere 0.01% of the world's oceans are protected as no-take marine reserves, and Russ's study makes a compelling case for increasing this proportion. It may seem premature to start popping corks based on the rebound of a single species, but coral trout are very important. They are top predators and as such, have tremendous pull on the Reef's biodiversity. Russ predicts that over time, the benefits reaped by the no-take zones will extend to other areas where fishing is allowed. Fish larvae do not stay in the same place and there is every chance that they will start to immigrate to unprotected areas and boost the populations there too. That is good for both fish and fisheries and it would hopefully mean that whatever fishing is allowed on the Reef would become sustainable.

The fishing bans have had more indirect benefits too; they have protected the corals from the predatory crown-of-thorns starfish (*Acanthaster planci*). The starfish is a voracious hunter of corals and a massive problem for reef conservationists. It is bad practice for any science writer to anthropomorphise an animal, but the crown-of-thorns really does look incredibly, well, evil. Its arms (and it can have as many as 20) are covered in sharp, venomous spines. As it crawls over the reef, it digests the underlying coral by extruding its stomach out through its underside. The starfish is a frequent pest but from time to time, its numbers swell into plagues of thousands that leave the dead, white skeletons of corals in their wake. These outbreaks eventually die off as the starfish eat themselves out of food supplies, but not before seeding downstream reefs with tiny larvae that drift along the southern currents. During their peak, they destroy far more coral than other disturbances such as bleaching events or hurricanes.

Hugh Sweatman at the Australian Institute of Marine Science found that these outbreaks are much less frequent in the no-take marine reserves.[38] Every year between 1994 and 2004, Sweatman carried out a census of starfish numbers in up to 137 areas across the Great Barrier Reef's massive length. It is possible that the protected fishes eat starfish, but the coral trout, whose populations have boomed in the no-take reserves, specialise in hunting fish not starfish. Sweatman thinks that the links between fishing and the crown-of-thorns are more complicated. His idea is that protecting the big fish like coral trout controls the numbers of smaller fish like wrasse. That puts less pressure on the populations of small invertebrates that in turn, feed on baby starfish. So, more big fish means fewer small fish, which means more invertebrates, which means fewer starfish. That certainly sounds convoluted, but other studies in recent years have provided examples of similar domino effects cascading down food chains. For example, the loss of big predators like sharks in the eastern seaboard of America destroyed local populations of scallops, clams and oysters, to the extent that local economies went bust. Sharks obviously do not feed on shellfish, but they do eat mid-level predators like rays and skates. Without the sharks to keep them in check, these intermediate hunters flourished and took out record numbers of animals at the bottom of the food chain. So Sweatman's theory has precedents elsewhere. It is also testable. Baby starfish live hidden in rubble after they settle in a new location and if Sweatman is right, the diversity of small invertebrates living among this rubble should differ between the no-take areas and the fished ones.

Regardless of the chain of events, it is clear that establishing no-take reserves is an effective way of stemming the crown-of-thorn's onslaughts. That is would be a massive coup for conservationists, especially since some of the richest and most threatened reefs on earth fall within the natural range of the crown-of-thorns. Any step that we can take to protect these most beautiful of ecosystems is one worth taking, and if the measures have other proven benefits – as the no-take reserves do – establishing them becomes even more compelling.

Uranium on the menu

The countryside around Iraq and the Balkans are still suffering from the ravages of wars fought in the 1990s. The environment is littered with the potentially dangerous remnants of military weapons – depleted uranium. Depleted uranium is what's left over after 'enrichment', when uranium-235 is separated from natural uranium. This isotope is suitable for nuclear reactors and weapons, and the remainder consists of uranium-238, a less radioactive isotope with a longer half-life. This "depleted uranium" is valued by the military for its high density and is often combined with titanium to produce an alloy used in both armour-piercing weapons and defensive plating. But penetrating rounds are not the only potential threat to human health posed by depleted uranium. The substance is still radioactive, can cause heavy metal poisoning, and burns spontaneously on impact to produce aerosols of uranium compounds. These potential risks have been downplayed by many reports but they make the use of depleted uranium in munitions highly controversial, especially when locals have to deal with the traces that litter war torn landscapes after battle has ceased.

Now, a new study shows that very unlikely allies may be helping to clean up these remains. Marina Fomina from the University of Dundee found that several species of fungi can not only thrive on depleted uranium, but also convert it into stable minerals.[39] Together with a team of British researchers, Fomina found that a large number of different species could happily colonise small wedges of depleted uranium. The fungi covered the wedges with large networks of long, branching cells called hyphae. Uranium wedges corrode naturally as they react with moisture in the environment to form uranium oxides, whose black and yellow hues were clearly visible. The tangles of fungal hyphae sped up this process by trapping even more water and pumping out hydrogen ions and other molecules that acidify the local environment. These conditions enabled the fungi to corrode the surface of the uranium fragments, which lost about 8% of their weight in a three month period. As a direct response to depleted uranium, the fungi also excreted organic acids such as oxalic acid that bind to uranium. It is a strategy that fungi also use to deal with other heavy metals and it converts uranium into a form that the fungi can take up. Indeed, some of the hyphae started turning yellow themselves, a sign that they had started incorporating the uranium salts into their network.

Amazingly, about 30-40% of the dry weight of the exposed fungi was made up of uranium. When Fomina looked at the fungi under the microscope, she found that the hyphae were encrusted by crystalline sheaths made of uranium minerals. The uranyl ions produced by the fungi's corrosive actions had reacted with phosphate ions released by the fungi themselves. These resulting uranium-phosphorus minerals, such as uramphite and chernikovite, formed large crystals that enveloped the hyphae. In these mineral guises, uranium is much more stable, and is effectively locked away for the foreseeable future. It cannot be taken up by plants and worm its way up the food chain. Fomina's study suggests that simple fungi could find themselves recruited into strategies designed to reclaim soil polluted by depleted uranium.

When ants break up with trees

The natural world is full of alliances forged between different species, cooperating for mutual rewards. The relationship between ants and acacia trees was one of the first of these to be thoroughly studied. But new research suggests that this lasting partnership may be sundered by the unlikeliest of reasons – the decline of Africa's large mammals. Acacias are under constant attack from hungry animals, from tiny caterpillars to towering giraffes. In response, many species like the whistling-thorn tree (*Acacia drepanolobium*) recruit colonies of ants as bodyguards. Any hungry herbivores eager to chomp on the acacia's leaves quickly get a mouthful of biting, stinging ants. The tree is a fair employer. In return for their services, its ant staff receives sugary and nutritious nectar as food, and hollow swollen thorns called 'domatia' as board. But this pact is a fragile one. Todd Palmer from the University of Florida and colleagues from the USA, Canada and Kenya found that it rapidly breaks down if the large animals that graze on the acacia disappear. Without the threat of chomping mouths, the trees reduce their investments in bodyguards to the detriment of both partners.

Palmer demonstrated this with plots of land in Kenya's Laikipia Plateau, where fences have kept out large plant-eaters for over a decade. Since 1995, no herbivore larger than a small antelope has entered the four-hectare "exclosures" in an attempt to study the effect of these animals on the local ecology. Within these 10 years, Palmer found that the majority of trees produced fewer domatia and less nectar and unexpectedly, the strongest

alliances were hit the hardest.[40] What were once happy partners quickly became selfish rivals.

Several species of ant compete for a place on the acacia's branches and the most abundant of these – *Crematogaster mimosae* – is also the most cooperative partner. It relies heavily on the domatia for shelter and aggressively protects the tree in return. But with no herbivores around, its services were not required and the trees started evicting it. The proportion of acacias colonised by *C.mimosae* fell by about a third. The partnership soured from both ends. The ants, with less food and smaller homes, became twice as likely to farm sap-sucking scale insects. Their waste fluid is a sugary liquid called honeydew that the ants drink, but to make it, they need to suck the juices of the tree. The ants also slacked off and were half as likely to marshal a defence against marauding plant-eaters. A second species *Crematogaster sjostedti* actually seemed to benefit from the trees' reduced investment. It is much less common than *C.mimosae* and takes a more relaxed attitude to the partnership. It could even be viewed as a parasite, for it defends the tree less aggressively and ignores the domatia, nesting instead in boreholes excavated by beetle larvae. As such, when the trees reduced their provisions, *C.sjostedti* was not fussed and with its competitor suffering, it more than doubled in abundance. A third species, *Crematogaster nigriceps* is even rarer and even more parasite than partner. Although it relies on domatia like *C.mimosae*, it also actively prunes any growing shoots that the tree sends out. This isolates it from the rest of the canopy, which prevents more competitive species like *C.mimosae* and

C.sjostedti from invading. But it also sterilises the tree. Despite this, trees harbouring *C.nigriceps* still produced the same sizes of nectaries and domatia even in the herbivore 'exclosures'. Palmer thinks that the trees cannot tell the difference between the munching of large herbivores and the ants' own pruning. Both send the same physiological signals that tell the tree it is being attacked, and that it should house defenders.

Palmer is keen to point out that the changes seen in the acacias are not evolutionary ones. The experiment's 10-year span is much shorter than the life of an acacia and with no passing generations, there is no clay for natural selection to sculpt with. The changes in domatia and nectar are part of a flexible program of behaviour enacted by the tree in response to changing pressures from plant-eaters. But in this case, it works to the acacia's disadvantage. You might expect that taking away large animals that feed on the trees should benefit them in the long run. In fact, the opposite is true. The key change is that the parasitic *C.sjostedti* replaces the cooperative *C.mimosae* as the dominant species in the ant-acacia community. And trees that bear this ant grow more slowly and are twice as likely to die as trees that harbour the others. Long-horned beetles are the problem. They are a destructive species that bore into the acacias' trunks. *C.mimosae* and *C.nigriceps* kill these pests, but *C.sjostedti* actually encourages them because it lives in the holes they leave behind. As the herbivores left and *C.sjostedti* rose to power, the beetles thrived and the trees paid the price.

Palmer's study elegantly shows that the fall of one species can ripple across an entire habitat in unforeseen ways. It is an increasingly common theme in modern ecology. Here, it sundered the collaboration between two cooperating species to the detriment of both, and left the door open for an invading parasite. While the situation in the study was created by scientists, it is being mirrored all over Africa. Several large plant-eating species are endangered or vulnerable, including the African elephant and both species of rhino. If they decline further, the entire ecosystem could feel the effects, from the tallest tree to the smallest ant.

Cottoning on to GM crops

Genetically modified crops have received a frosty welcome in the UK, and more widely in Europe. Those opposed to such crops worry, among other things, that they could affect the surrounding plant-life by simply outcompeting them or by spreading their altered genes in a round of genetic pass-the-parcel. But Kong-Ming Wu from the Chinese Academy of Agricultural Sciences found that genetically modified cotton can indeed affect surrounding plants – in a positive way.[41] The altered cotton is designed to kill a very hungry caterpillar but it can also protect other species of plants from chomping jaws. In doing so, this "Bt cotton" could help to reduce the need and demand for other sprayed insecticides.

Bt cotton has been loaded with insect-killing genes taken from a bacterium called *Bacillus thuringiensis* (hence "Bt"). This species lives in soil and the surface of plants, and it produces crystals of proteins that are toxic to hungry insects. If they are swallowed, they stick to molecules in the pest's gut, breaking down its lining and allowing both *B.thuringiensis* spores and colonies of normal gut bacteria to invade. This wanton spread of bacteria is what kills the insect. For decades, the protein crystals have been sprayed over crops as a pesticide. The chemicals degrade easily and pose no risk of contaminating groundwater. They are also highly specific - toxins from a specific strain of Bt will usually only affect a few species of insect, so insecticides can be very finely tailored to control certain pests while minimising collateral damage to other insects, and indeed other animals. In Bt cotton, these genes have been implanted so that the plant itself produces the relevant insecticides without the bacteria. By relocating these defences to the plant's own genome, scientists can greatly reduce the need for spray-on pesticides, and the power of this 'transgenic' technology has made Bt cotton a favourite of farmers from developing countries. Today, it is one of the most common of genetically modified crops and a third of the world's cotton-growing land is taken up by these Bt varieties.

In China, the transgenic cotton has proved to be an important defence against a type of moth called the cotton bollworm (*Helicoverpa armigera*). The larvae of this destructive pest has wide-ranging tastes and it will eat cotton as well as wheat, corn, peanuts, soybeans and vegetables. The bollworm launches at least four waves of attacks on Chinese crops ever year, it migrates over long distances and its outbreaks can span several provinces. Wu and colleagues have been monitoring these outbreaks since 1992 in six different Northern Chinese provinces. Together, the team investigates over 38 million hectares of farmland cultivated by 10 million or so farmers. At first, the populations of the bollworm found on local plants was very high, but all that changed in 1997, when Bt cotton was first approved for commercial use. In the years since then, Wu found that both the density of bollworm on cotton plants and the density of its larvae on other host crops had fallen. Wu also looked at rainfall and temperatures over the 15-year span and found that neither of these could explain the falling bollworm census. Only the rising use of Bt cotton did that and it extended an umbrella of protection over all the local plants.

Wu thinks that cotton serves as the bollworm's main staging ground, the site where the first generation of bollworms to invade an area laid their eggs. But Bt cotton kills any larvae that hatch from this first wave of eggs and acts as a botanical dead-end for the moth. The entire crop becomes a trap. Even so, there is still a worrying chance that the insects will simply evolve resistance to the insecticides. The possibility of this happening increases if farmers continuously plant the same variety of cotton, carrying the same Bt toxins. In the US, the solution (which farmers must adopt by law) is to plant Bt cotton alongside normal varieties that lack any sort of genetic modifications. These normal plants act as refuges for any invading pests and keep them contained while relaxing any evolutionary pressures to develop resistance. The situation in China is a bit more difficult because the farming population is far greater, and the prospect of educating them all about this "refuge management" technique is too daunting. Thankfully, the farmers have developed an equivalent method almost by accident; they plant a range of different crops alongside cotton, including soybean, corn and peanuts. The Bt cotton provides some shielding from bollworms but these other plants, which are not individually defended, act as a refuge.

It's worth noting that *Bacillus thuringiensis*, which has donated its genes to Bt-cotton, is one of the few insecticides whose use is allowed within the definitions of organic farming. The fact that the bacterium is found "naturally" in the soil apparently elevates it to a high enough status that its proteins can be liberally sprayed over organic crops. The genetically modified cotton, of course, uses exactly the same proteins to ward off the mandibles of insects; the only difference is that the capacity to make these proteins has been relocated to the plant's own genome rather than that of the bacteria. Wu claims that despite its benefits, Bt cotton must be considered as just part of the solution for controlling China's agricultural pests. In some parts of the country, plants are being attacked by sap-sucking bugs because the introduction of Bt cotton has meant that farmers are spraying fewer insecticides with wider ranges of targets.

Farming with fire

Imagine that you have been given responsibility over a tract of land. Your goal is to prevent the local habitats from becoming degraded and to maintain their precious biodiversity (increasing it, if at all possible). At the same time, you have to find a way to eke out a living. Of the many possible ways of doing this, regularly and deliberately setting fire to the local plants might seem low on the list, but that is exactly what Aborigine populations in Australia have been doing for centuries. Now, we have evidence that this seemingly counter-intuitive strategy does indeed work.

A team of American anthropologists led by Rebecca Bird at Stanford University studied the practice of "fire-stick farming" among the Martu people of Australia's Western Desert.[42] The Martu live mostly as hunter-gatherers and supplement their food with the odd supply bought from local outstations. Their homelands are mostly dominated by sandy plains and the ubiquitous spinifex grass (*Triodia*), and these are the areas that the Martu start fires in. They have different words for land at various stages of post-fire recovery: *nyurma* is freshly scorched earth, *waru-waru* describes land where shoots have started to sprout; *mukura* turns up after a few years when grasses, flowering shrubs and edible plants have arrived; *mangul* occurs a few years later still when the growing spinifex starts to outcompete edible plants, leading to *kunarka* when the spinifex starts to die and leaves behind sterile hollows. These "successional stages" follow one after another in predictable ways and the Martu only ever set fire to the last two, when spinifex is dominant. In doing so, they effectively press an ecological reset button, allowing plants to return to areas that had previously been won by the unbeatable spinifex.

The burning process is an important social ritual that the Martu take great pride in and it is far from haphazard; fires are carefully ignited in areas that are upwind of known firebreaks and they are typically set during the winter season when winds are stronger and more consistent in direction. With such control, most fires burn out very quickly and it is in an individual's interest to ensure that they do so; everyone has the right to start a fire, but individual fire-starters are accountable for any that rage out of control and damage sacred sites. The women do the majority of the burning and they do so to hunt for goannas (monitor lizards), pythons and skinks, which typically hide in spinifex bushes and are hard to spot. They flee from flames by digging fresh burrows and by searching for fresh tracks in the wake of the blaze, hunters can extract the lizards with a specialised digging stick. The Martu are all too aware of the comparative ease of goanna-hunting on freshly burnt ground, but they cite many other benefits too - the practice provides more food for both people and animals (particularly large ones like kangaroos and emus) and somewhat paradoxically, it can also prevent big bushfires.

Bird's team evaluated these claims by using satellite images to analyse 34 circles of land in the Western Desert, each with a radius of 3km. They found that those which the Martu frequently set alight had a much greater degree of plant diversity than those where fires only ever erupted spontaneously due to lightning strikes. They contained more "edges", where one type of habitat gives way to another. And they contained a broader and more even range of vegetation at different successional stages, from the early collection of shoots (*waru-waru*) to the later kingdoms of spinifex (*mangul*). So by carefully scorching the land, the Martu actually *protect* it from habitat loss and increase its biodiversity. They use fire not to destroy their environment but to rearrange it, creating a mosaic of different communities of plants at a local level. They burn the land to construct their own ecosystem in a way that directly benefits them.

In 2002, Bird followed more than 150 hunting excursions and found that after burning, the hunters are able to catch about 25% more food, equivalent to an extra 400 calories per day. They also take less time to net this bounty and the chance of failure on any individual hunt falls from 22% to 4%. Clearing away the mature spinifex removes the hiding places that small reptiles shelter in but it also creates a finer mosaic of different habitats that are easier to move about it, which means that hunters spend less time making their way through difficult terrain. These benefits also fall dramatically as habitats become more diverse, which makes burning a self-limiting activity. After a certain point, there is no extra worth in starting more fires, and the Martu know it.

The Martu's man-made fires also help to control natural ones. Fires started by lightning strikes often sweep through vast tracts of bush and find plenty of fuel in the large areas of spinifex. But in the patchier mosaics created by the Martu, natural bushfires burn out more quickly. That has two paradoxical consequences: the Martu do not actually subject the land to any more burning than it would normally experience, even though they actively start fires; and the spinifex actually benefit from being periodically burnt because they are less likely to be completely wiped out from local areas by large-scale bushfires.

Bird's study suggests that fire-starting, as the Martu assert, is a very effective way of managing their land. The hunter-gatherers and the local species benefit from it, the land becomes more diverse, bushfires are kept in check and most importantly, it is a self-limiting strategy with no incentives for overuse. If any party gets a raw deal, it is probably the goannas...

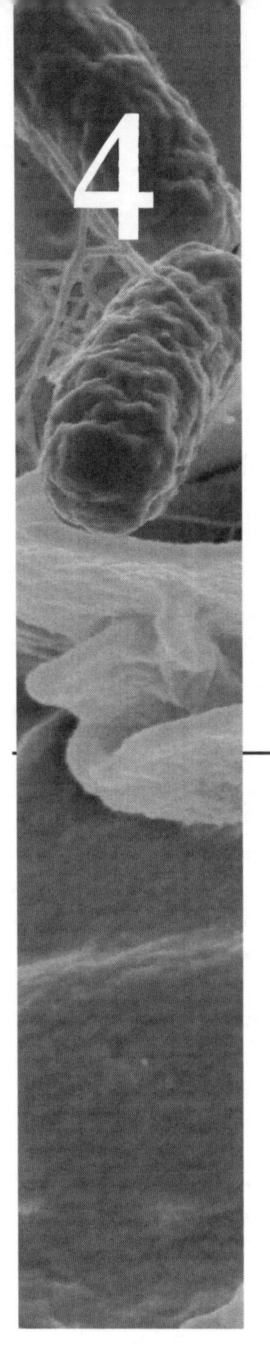

Hidden cultures

Influenza's world-tour, snow-making bacteria and the world's first meta-virus

Walking in a bacterial wonderland

The next time you watch a snowfall, just think that some of the falling flakes house bacteria at their core. It is a well-known fact that water freezes at 0°C, but it only does so of its own accord at much colder temperatures of -40°C or colder. At higher temperatures, it needs help and relies on microscopic particles to provide a core around which water molecules can clump and crystallise. These particles act as seeds for condensation and they are rather dramatically known as "ice nucleators". Dust and soot are reasonable ice nucleators but they are completely surpassed by bacteria, which can kick-start the freezing process at higher temperatures of around -2°C.

Ice-forming bacteria like *Pseudomonas syringae* rely on a unique protein that studs their surfaces. Appropriately known as ice-nucleating protein, its structure mimics the surface of an ice crystal. This structure acts as a three-dimensional template that forces neighbouring water molecules into a pattern matching that of an ice lattice. By shepherding the molecules into just the right place, the protein makes it easy for them to bond together in the right conformation, and greatly lowers the amount of energy needed for ice crystals to start growing. This ability of bacteria to seed the growth of ice crystals has been known for some 40 years but it is taken till now to show just how widespread these living ice-makers are.

Brent Christner from Louisiana State University did it by collecting large samples of snow from 19 fresh snowfalls in regions as diverse as the USA, France, and Antarctica, and isolated the ice nucleators by filtering the melted snow.[43] Surprisingly, he found traces of cells containing DNA among the nucleators in *every single site.* Both heat, which deforms the ice-nucleating protein, and lysozyme, an enzyme that wrecks a bacterium's outer wall, robbed the vast majority of these particles of their ability to form ice. Clearly, our atmosphere is rife with ice-forming microbes. They may even have an important role in the genesis of clouds, especially at relatively warm conditions above -7°C, when ice does not form spontaneously.

Most of the bacterial ice nucleators infect plants and Christner thinks that they may have been swept away from the surface of infected hosts by gusts of wind. Once airborne, the bacteria can be transported across remarkable distances. Certainly, Antarctica is not known for its greenery and the bacteria that Christner found in snowfall from this region must have travelled from other continents. Over time, the floating bacteria become surrounded by ever-expanding ice crystals and eventually fall to the ground as snow or rain. On the way, they could potentially hit new plants to infect. In this way, you could almost think of the atmosphere as a giant

conveyor belt that provides bacteria with a free lift from one plant to another. It could be just another stage in the infection cycle and bacteria could be using clouds for transport, just like malaria parasites use mosquitoes to jump between hosts. The ice-nucleating protein may well have evolved for this very purpose. By encouraging the growth of ice crystals, it increases the chances that a bacterium will fall back to earth, in the direction of new hosts.

Sputnik – the virus of viruses

Viruses may cause disease but some can fall ill themselves. In 2008, Bernard La Scola and Christelle Desnues at the University of the Mediterranean discovered the first virus that targets other viruses, which they playfully named Sputnik, after the Russian for "fellow traveller". It is so unique that they have classified it in an entirely new family – the "virophages".[44] The name takes after another group of viruses called bacteriophages, which infect bacteria. The story of Sputnik started in 1992 with some dirty English water. A group of scientists were studying a species of amoeba taken from a cooling tower in Bradford, England, when they discovered a microscopic giant – a virus so large that it was originally mistaken for a bacterium. It was only in 2003 that La Scola and colleagues conclusively showed that the new find was indeed a virus. But *what* a virus – APMV, or 'mimivirus', measures a whopping 400 nanometres across, dwarfing other types like the common cold.

The search for giant viruses continued. La Scola's team identified another strain of APMV by inoculating the same species of amoeba with water taken from another cooling tower, this time a Parisian one. The new specimen eclipsed even the original giant in size, and the researchers (who clearly have a sense of humour) decided to call it 'mamavirus'. When this record-breaker infects amoebae, it forms gigantic viral factories that pump out new copies of itself. When the team looked at these under an electron microscope, they found an even bigger surprise. They saw the equivalent of microscopic Russian dolls – tiny viral particles, just 50 nanometres in size and distinct from mamavirus itself. Sputnik had landed.

La Scola and Desnues found that Sputnik could not multiply within the amoeba by itself; it could only spread within cells that had also been infected with mamavirus. But Sputnik is no partner – by hijacking the mamavirus's replication machinery, it spreads at the expense of its larger host and substantially hinders its reproduction. In the presence of the tiny intruder, mamavirus particles assemble abnormally and surround themselves with unusually thick outer coats. As a result, their ability to infect the amoeba fell by 70%. This lifestyle means that the virophage name is perhaps a bit misleading. Bacteriophages reproduce within the cells of bacteria, whereas Sputnik is a satellite virus, in more than name only. Like hepatitis D, it depends on another virus coinfecting a host in order to spread. But it is the fact that it does so at the expense of the mamavirus that makes it a true parasite.

In comparison to its sizeable host, Sputnik is tiny and sports a genome that is almost a hundred times smaller. Its 18,000 base-pairs of DNA contains just 21 genes and when La Scola and Desnues analysed these, they found that Sputnik is a genetic chimera – a mish-mash of different genes from different sources. Thirteen of these have no equivalent in any other known virus, while the remainder have similarities to genes from other viruses, bacteria and even more complex cells. Three of these are closely related to mamavirus genes, suggesting that this tiniest of parasites has been raiding genetic material from its host and from other viruses. La Scola and Desnues even suggest that Sputnik could be acting as a genetic mule, shuffling genes between giant viruses. It could even explain why mimivirus has mysteriously and recently picked up bacterial genes of unknown origin.

The fact that mamaviruses can "get sick" themselves is further evidence that viruses are indeed living things. But Sputnik's discovery has far larger implications – it shows that we have barely begun to scratch the surface of virus diversity. This unique microbe was found by a series of fortuitous circumstances and who knows how many others like it are out there waiting to be discovered. Until now, no one had any idea that virophages even existed and that is the true beauty of scientific discoveries like these – they reveal to us the true depths of our ignorance.

The flu virus's world-tour

Of the different types of flu virus, influenza A poses the greatest threat to human health and at any point in time, about 5-15% of the world's entire population are infected with these strains. Together, they kill up to half a million people every year and the death toll rises sharply when pandemics sweep the globe. This year, two papers shed new light on the origins of these epidemics. By prying into the private lives of flu viruses, the studies provide fresh clues about the birthplaces of new strains, their flight plans around the world and the locations of possible 'viral graveyards'. The findings could help health organisations to design better strategies for monitoring the emergence of new strains and selecting vaccines that will do the most good.

All influenza A viruses are deceptively simple. They all contain eight gene segments that provide coded instructions for building just 11 proteins. Two of these – haemagglutinin and neuraminidase – are of particular note, for they help the virus stick to a host cell and are used by scientists to distinguish between different types. The H1N1 subtype, for example, was responsible for the 1918 'Spanish flu' pandemic, and H5N1 is the avian flu subtype that so concerns us today. However, the dominant subtype is currently H3N2, which first found its way into humans in 1968 in a form that has since become known as 'Hong Kong flu'. Understanding how, when and where these strains of flu evolve is a complicated business, especially because existing strains are liable to mix and match to form new ones. This process, known as 'reassortment', often happens when viruses of different types infect the same cell and swap gene segments between them. This game of genetic swapsies makes it very hard to draw a simple family tree that unites different types and strains of flu.

To understand the evolution of influenza subtypes on a large scale, Andrew Rambaut from the University of Edinburgh, together with an international team, analysed over 1,300 samples of H1N1 and H3N2 collected across a 12-year period from New York State and New Zealand.[45] It is well-known that both subtypes become more common during the

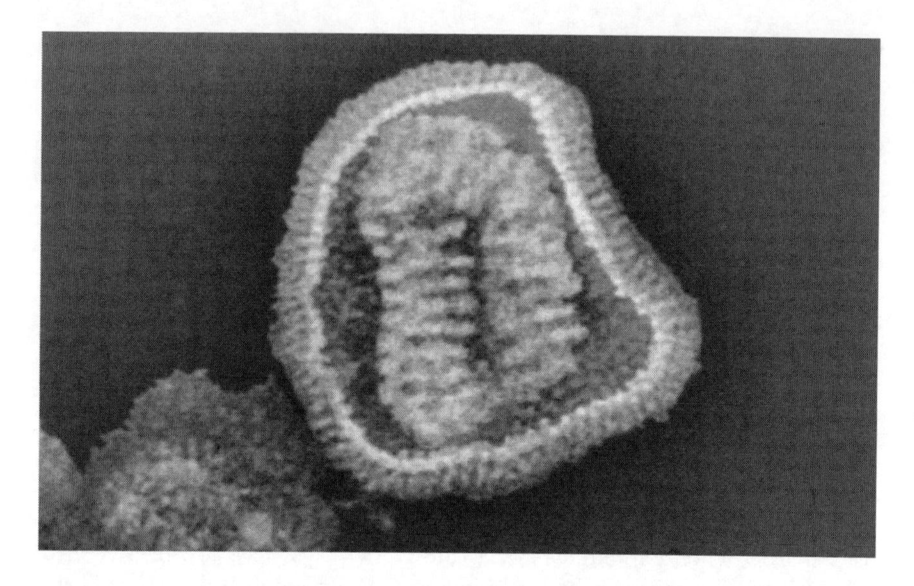

winter months, but the forces that drive these seasonal cycles are unclear. Rambaut found that at the start of each winter season, there is a sharp rise in the genetic diversity of H3N2 viruses, followed by an equally stark bottleneck at winter's end. As expected, the timing of these evolutionary peaks in the Southern Hemisphere is offset from those in the North by six months; after all, winter in Australia happens during summer in Britain. The patterns of diversity in the two subtypes are also intertwined. H1N1's genetic diversity only peaks strongly in years when that of H3N2 peaks weakly, suggesting that the dominant virus either causes a herd immunity that protects people against its rival, or simply outcompetes it.

Rambaut suggested two possible explanations for the fact that viral diversity repeatedly peaks in winter epidemics, despite going through summer lulls. The first is that chains of infection manage to linger between epidemics as people pass the virus to each other. These persistent chains would cause blooms of diversity every winter as the viral populations start to rise. But it is an unlikely theory. Every year, there is a good chance that random events will cause strains to go extinct (say, if they fail to find a new host in time), which makes their continued survival across the years very improbable. The second theory is that viruses 'import' genetic diversity from more stable reservoirs in other parts of the world, and this is better supported by genetic evidence. Typically, family trees of flu samples show that viruses taken from different seasons in the same location show little direct relationships to one another. That argues against the idea that strains persist at a local level, and suggests that they are instead 'seeded' by a source somewhere else in the world. But where?

Rambaut proposes that the tropics are the source. They act as a large melting pot for influenza evolution, where new strains emerge and are exported to temperate regions, where they fuel seasonal epidemics. In these tropical 'sources', the populations of both human and animal hosts is very large and infections can spread indefinitely all year round. As such, natural selection can shape the evolution of new strains more efficiently than in the temperate 'sinks' with their seasonal bottlenecks. This theory could explain other puzzling characteristics of influenza evolution, like the fact that the haemagglutinin protein often changes substantially between epidemics but changes very little within each separate one. Rambaut writes that it is possible that tropical regions generally represent ideal source populations for flu viruses and that it will be necessary to study more samples from tropical regions to test this idea. As luck would have it, a second group did just that and they managed to pinpoint the crucible of flu evolution even more precisely.

Colin Russell from the University of Cambridge, heading another international group, focused on the virus's haemagglutinin protein (HA), the main target of our own immune systems. His team followed the worldwide evolution of H3N2 strains by analysing the immune reaction triggered by 13,000 samples, taken from six continents over a five-year period.[46] The team found that during this time, newly emerged strains showed up in East and South-East Asian countries about six to nine months earlier than they did in other parts of the world. On the other hand, the viruses' travels to South America were much delayed and they took a further six to nine months to reach it. The researchers also sequenced part of the HA protein in about 10% of their samples and the genetic relationships between them confirmed that new strains follow the same east-to-west route. As before, this pattern can be explained in two ways involving local persistence or external seeding. The local persistence theory is more complicated, for it suggests that new H3N2 strains emerge independently around the world as a result of similar evolutionary pressures. Like Rambaut, Russell denounced this idea and he did so by sequencing HA samples to build a virus family tree, showing how different strains were related to one another. The tree revealed that strains were always descended from those hailing from other parts of the world, rather than those from earlier local epidemics.

Like Rambaut, Russell agrees that new viruses are 'seeded' from a specific source, but his data contradict Rambaut's suggestion that this source is the entire tropical belt. For a start, South America always carried flu strains that were less advanced than those of Asia, and it never gave rise to its own independent flu lineages. Instead, Russell claims that the vast majority of new H3N2 strains originate in East and South-East Asia before spreading worldwide, and this theory is supported by the HA family tree.

The 'trunk' of the tree is dominated by strains from East and South-East Asia and strains from temperate regions branch off from these oriental predecessors. However, it is impossible to narrow things down even further. Some researchers have suggested that Southern China is the sole source of new flu strains but Russell found that no single country in this part of the world consistently shows more advanced viruses than those of their neighbours. New strains often emerge in countries other than China, although for some reason, Malaysia, Thailand and Japan contribute less than their expected share of new strains.

Russell suggests that this part of the world seeds the planet's flu population because the viruses find it easy to circulate continually. East and South-East Asian countries have very variable climates and go through their rainy seasons (when influenza infections are most common) at different points of the year. So these countries go through epidemic cycles just as more temperate regions do, but the timings of their peaks and troughs are very different. As a result, flu viruses can jump from one country's epidemic to another, creating a continuous network of infection. Russell found that the viruses within this network are fairly cliquey and are almost never seeded by strains coming in from other parts of the world. Instead, this network exports new viruses to other countries, a job made easier by Asia's strong trade connections to Europe, North America and Australasia. It seems that the region ships flu as readily as it does electronic goods. South America, on the other hand, has little trade with East and South-East Asia, which may explain why it lags behind so much in terms of flu evolution. To get there, viruses need to make a stopover in North America and Europe first. In fact, South America as a whole may represent an evolutionary dead-end for H3N2, a sort of viral graveyard. Africa is the only remaining wild-card; Russell admitted that there is not enough data from that continent to rule it out as a potential source of flu.

Knowing how flu viruses evolve is one thing, but the real challenge is to be able to predict these patterns in the future. Russell's work was supported by the WHO's Global Influenza Surveillance Network, which monitors the world's influenza population and meets twice a year to decide which strains from three subtypes (including H3N2) will be included in the flu vaccine for the next season. Russell recommends that the WHO keep the East and South-East Asian flu network under increased scrutiny. Close inspections of this influenza hotspot will help us to detect new strains as they emerge and determine which are most likely to seed the rest of the world. While the existing vaccine is very successful, a better understanding of our enemy will allow us to stockpile even more appropriate weapons. These initiatives will be even more effective if they pay attention to the entire influenza genome, as Rambaut did, rather than the HA protein in isolation.

Size does matter

They say that size does not matter, but try telling that to bacteria. Most are very small and there is a reason for that – they rely heavily on passive diffusion to ferry important nutrients and molecules across their membranes. To ensure that this happens quickly enough, bacteria need a large surface area relative to their volume – if they become too big, they will not be able to import enough nutrients to support their extra size. These constraints greatly limit the size of bacteria. The larger ones solve the problem by being extremely long and slender, or by using an internal compartment called a vacuole to push their insides to their outer edges. But even these solutions have their limits, and the simple fact is that most bacterial cells are far smaller than those of the more complex eukaryotes – a group that includes all known animals, plants and fungi. But one group of bacteria called *Epulopiscium* flouts this rule. These species have developed a genetic trick that allows them to grow to gigantic proportions (well, for bacteria anyway). While a typical bacterium like *Escherichia coli* is a mere two micrometres long, *Epulopiscium* can grow up to a length of 300 micrometres. That is not much smaller than the full stop at the end of this sentence, making *Epulopiscium* a bacterium that you can actually see with your naked eye. It is certainly comparable in size to most eukaryotic cells. And the secret to its mammoth dimensions? DNA and lots of it.

Jennifer Mendell from Cornell University studied the *Epulopiscium* by working in the lab of Esther Angert, who first identified it as a bacterium in 1993. Angert's lab had previously shown that the colossal microbe contains a large quantity of DNA concentrated near its membrane, and Mendell wanted to work out how much. But doing that was not easy. When it comes to choice of home, *Epulopiscium* redefines the word 'picky', for it only lives in the intestines of surgeonfish. Different species colonise different fish and Mendell worked with one that lives only in the guts of the bulbnose unicornfish. While most microbiologists can culture their subjects, Mendell had to collect hers by spearfishing in the Great Barrier Reef! Her hard work paid off and she speared enough samples to be able to measure the amount of DNA in over 10,000 *Epulopiscium* cells.[47]

Most bacteria reproduce by splitting in two or budding off daughters but *Epulopiscium* produces offspring in a way that is a lot like a live birth. It grows two daughters inside itself, until they fill the mother cell entirely and burst out of it. These larger, mature cells contained about 250 picograms of DNA, while the smaller, immature ones contained about 85 picograms. That is still a huge amount – a typical human cell only contains about six picograms. It is unlikely that all of that DNA belongs to a single, massive genome because only a small amount is passed onto each daughter. *Epulopiscium* may be enormous but its genome is fairly standard in size for bacteria – it just has many copies of it. Mendell demonstrated this by measuring the frequency of three bacterial genes that are always found once per genome. She found that the large cells contained an average of 41,000 copies of each of these genes for every 156pg of DNA, implying that every cell contained tens of thousands of genome facsimiles! This strategy of having more than two copies of the same genes (as humans do) is known as "polyploidy".

What use could the bacterium have for so many copies of its genome? Mendell offers several answers. For a start, the multiple copies allow it to tolerate genetic mutations that would kill most other bacteria – if something goes wrong with its DNA, it always has spare copies kicking around. Having more genetic material also allows it to become much larger, and being big has its advantages. The bacterium can travel to parts of the gut that are richest in food, and it is so large that predatory cells that would eat lesser bacteria find it too large to swallow. *Epulopiscium*'s lifestyle also provides a clue to the reason behind its genetic surfeit. It is a symbiont, a bacterium that lives in harmony with its fishy hosts. In their intestines, it finds a constant stream of warm, nutritious fluid and in return, it pays its landlord by secreting valuable molecules. If this molecular rent provides the fish with a survival advantage, the bacterium benefits too and will experience a strong evolutionary pressure to provide its host with even more. And indeed it does – it produces very high numbers of molecular pumps that shuttle proteins out of its borders for the benefit of its host. The extra genetic material helps it to do that, and Mendell thinks it is no coincidence that many other symbionts also carry duplicated genomes (albeit not to the same extent). The bacterium's extra genetic material is dotted around its edges, near the cell membrane. This means that new proteins can be created and directly fed into the membrane without any transport problems, in the same way that a retailer can minimise shipping delays by having multiple warehouses dotted around a region rather than a single central depot. This division of labour allows a single *Epulopiscium* cell can act as a colony with different parts responding independently to local conditions. Many of these advantages – large size, mobility, and defence against predators – are typically found in large eukaryotic cells or even multi-celled organisms but by accumulating thousands of copies of its genomes and redistributing them effectively, *Epulopiscium* has achieved all of these benefits through a completely different route. It is another testament to the wondrous diversity and abilities of bacteria.

Eating antibiotics for breakfast

Antibiotics are meant to kill bacteria, so it might be disheartening to learn that some bacteria can literally eat antibiotics for breakfast. In fact, some species can thrive quite happily on nothing *but* antibiotics, even at high concentrations. The rise of these drug-resistant bacteria poses a significant threat to public health and many dangerous bugs seem to be developing resistance at an alarming rate. The headline-grabbing MRSA may be getting piggybacks from livestock to humans, while several strains of tuberculosis are virtually untreatable by standard drugs. Worryingly, antibiotic resistance may be much more widespread than anyone had previously appreciated and some bacteria can even feast on the molecules that are meant to kill them. Gautam Dantas from Harvard Medical School managed to culture antibiotic-eating bacteria from every one of 11 soil samples, taken from farmland and urban areas across the US.[48] All eleven were positively loaded with a diverse group of bacteria that were extremely resistant to a wide range of antibiotics at high concentrations. In their natural environment, these soil bacteria are frequently exposed to a massive array of natural antibiotics secreted by plants and other microbes, and they have evolved ways of detecting and evading these molecules. These resistant strains act as a living reservoir of innovative genetic defences, known as the 'antibiotic resistome'.

Dantas searched for resistant bacteria by culturing colonies that could grow in solutions where antibiotics were their only source of carbon. He tested 18 different antibiotics that are used to kill a variety of different bacterial species. Some of these were natural, others man-made; some were old, others new. But every single one managed to support at least one strain of bacteria. Six of them, including commonly used drugs like penicillin, vancomycin, ciprofloxacin and carbenicillin, even managed to feed bacteria from all 11 soils. The degree of resistance in the soil bacteria was nothing short of extraordinary. Dantas cultured a representative set of 75 resistant strains and found that on average, they resisted 17 of the 18 antibiotics at low concentrations of 20 milligrams per litre. But even at higher concentrations of one gram per litre, each strain managed to stand firm against an average of 14 out of 18 drugs. When Dantas studied some of these strains more closely, he found that they nullified the drugs using similar techniques to the drug-resistant versions of disease-causing bacteria. Some shunted the antibiotics out of their cells with molecular pumps, others used enzymes to cut up the drugs, and yet others reprogrammed their own genetic code to deprive antibiotics of their targets.

The real danger is that these mostly harmless soil-living species could provide new defences that more dangerous ones can draw on to shrug off our best drugs. Bacteria are capable of passing genetic material between one another as easily as two humans might swap business cards, making it trivial for the soil-based super-bugs to pass their crucial genes on to more dangerous species. There is precedence for this already – last year, scientists found that drug-resistant versions of the deadly bacterium that causes bubonic plague have actually picked up their resistance genes from the common food poisoning bug *Salmonella.* In principle, bacteria should be particularly capable of successfully taking up resistance genes from closely related species. It is worrying then that Dantas's antibiotic-eaters belonged to such diverse groups. By establishing a family tree of the different strains, he found that they were members of at least 11 different bacterial groups, although over half of them came from just two orders – the *Burkholderiales* and the *Pseudomonadales.* These two groups include a wide variety of species that are known to infect hospital patients with weakened immune systems. They are known for their large genome sizes (well, large for bacteria anyway) and some scientists have suggested that these sizeable genomes allow them to metabolise a wide range of chemicals, antibiotics included.

The ability to eat antibiotics will come as no surprise to those who study bacteria for a living. Bacteria can colonise some of the most extreme environments on the planet and can survive on the most unlikely to food sources, from crude oil to toxic waste. Now, it seems that they can also survive solely on chemicals that are meant to kill them.

Taming Ebola

In a list of the most dangerous jobs in the world, 'Ebola researcher' must surely rank near the top. But if new research is anything to go by, it may soon fall several places. An international team of scientists have recently found a way to neuter the virus, making it simpler to study without risking your life. The altered virus looks like Ebola and behaves like Ebola, but it cannot kill like Ebola. It should make studying the virus easier and most importantly, safer.

The Ebolaviruses (*opposite page, top*) and their cousins, the closely related Marburg family, have a chilling and deserved reputation. In some outbreaks, 90% of those infected die from massive blood loss. There is no approved antiviral treatment. There is no vaccine. And given that it is almost a *rite du passage* for infectious disease scientists to contract the contagion they study,

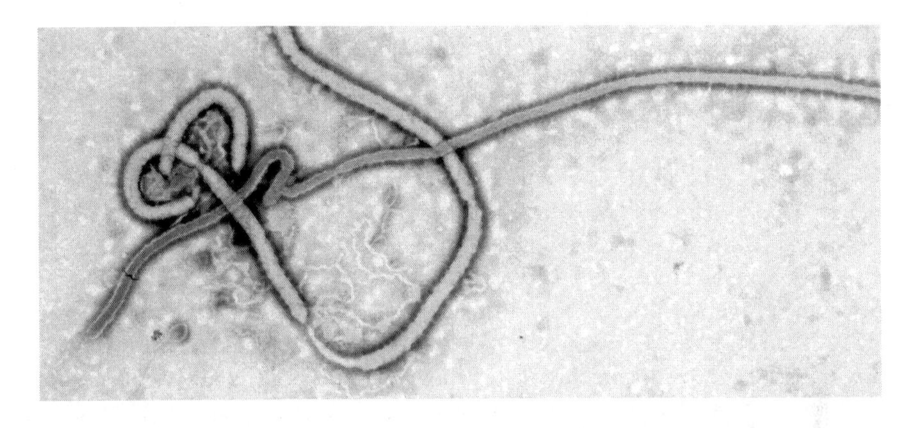

working with Ebola is a delicate affair. As such, Ebola research requires the highest level of safety possible – the Biosafety Level-4 laboratory. The stand-alone facilities are designed to be easily sealed and impervious to animals and insects. All routes in and out, including all pipes and ventilation, are peppered with multiple airlocks, showers and rooms designed to prevent any chance of escaping viruses. There are very few people who are qualified to work in such a prohibitive environment and those that do must wear a Hazmat suit at all times and breathe from a self-contained oxygen supply. No wonder then that the majority of Ebola research does not actually use live, infectious viruses. Scientists must instead settle for isolated proteins, proteins shoved into other, less harmful viruses, or even "virus-like particles". But these artificial systems are different to the virus proper, and using them is like staring at a complex machine through a cobweb-covered keyhole. Peter Halfmann from the University of Wisconsin has found a way around this, opening the door for scientists to get a proper look at the virus.

Ebola has a tiny genome that contains eight genes, which in turn provide the instructions for eight essential proteins. One of these – VP30 – is used by the virus to switch on its other genes and to make copies of itself. Together with colleagues from the US, Canada and Japan, Halfmann cut out VP30 from the Ebola genome and replaced it with a gene for an antibiotic called neomycin.[49] Without the gene, the virus is effectively castrated; it cannot carry out its normal deadly activities and it certainly cannot reproduce and spread to infect other cells. To grow at all, it needs to be placed in a specially created culture of monkey cells that produce the missing VP30. These cells allow the virus to 'come alive' but they also act like a biological prison, confining Ebola using its own need for the all-important VP30. The virus can happily thrive in cultures of these specially designed cells but it can never spread and infect normal ones. Under the electron microscope, the edited viruses were completely indistinguishable from their wild relatives in both size and shape. In the VP30 cells, they grew at a similar pace and produced their entire repertoire of proteins. But in normal cells, they completely failed to grow, and the team found not a trace of viral proteins.

There is, of course, a risk that the neutered viruses could somehow regain VP30 from the cells around them and regain their full infectious powers. Halfmann recognised this and carefully checked that the viruses were genetically stable. He was reassured – even after seven rounds of infection in the VP30 cells, the viruses retained the inserted neomycin gene and still failed to infect normal cells. The implications for Ebola researchers are immense. Now, they have a virus that can be studied outside the prohibitive (and prohibitively expensive) confines of a Biosafety Level-4 lab. They could replace VP30 with genes of their choice, such as GFP, a gene that produces a green, glowing protein that would make the virus easy to monitor. They could use Halfmann's viruses to study Ebola's life cycle in detail or screen thousands of potential anti-viral chemicals for new vaccines or treatments, in a way that is impossible with current artificial systems. When it comes to Ebola research, accept no substitutes.

An ecosystem of one

Most of the planet's ecosystems are made of a multitude of different species, rich tangles of living things all interacting, competing and cooperating in order to eke out an existence. But not always – in South Africa, within the darkness of a gold mine, there is an ecosystem that consists of a single species, a type of bacteria that is the only thing alive in the hot, oxygen-less depths. It is an ecosystem of one, living in complete isolation from the Sun's energy.

This incredible and unique habitat was discovered by Dylan Chivian from the Lawrence Berkeley National Laboratory, leading a large team of scientists from 15 institutes.[50] The group was interested in studying extremophiles, species of bacteria that live in the planet's harshest conditions – in this case, the rocks of the Earth's crust. At depths of a kilometre or more, bacteria face unique challenges that their counterparts on land or sea do not, including high 60°C temperatures and a lack of sunlight, oxygen and nutrients. The species that can beat these challenges are interesting to biologists because they provide insights into how life can persist on the edge of existence and, potentially, on other planets. To find such species, Chivian's team of bacteria-hunters did their work in a series of mines in South Africa's Witwatersrand basin. In one venture, they collected over 5,000 litres of water that had pooled in cracks in the rock almost three kilometres deep. They used a technique called 'metagenomics' to extract and analyse all the DNA from the sample. Metagenomics allows scientists to study the hidden worlds within any given habitat and to identify the multitude of bacteria and other micro-organisms that live there. Usually, the technique gives an interesting overview but would never be able to provide a complete census of the local species. But in the deep water, Chivian found a surprise. All the DNA belonged to a single species of bacteria, which they named *Candidatus Desulforudis audaxviator.*

The name is an apt one; "audaxviator" means "bold traveller" and comes from a quote from Jules Verne's novel *Journey to the Centre of the Earth.* The species' genes have been isolated from other crust samples before but this is the first time that it has been found in total isolation. And to a large extent, this community of one shares a single genome. The population has a staggeringly low level of genetic diversity; among the 2.3 million base pairs that make up its genome, only 32 displayed mutations that cropped up more than once. *D.audaxviator* is not one for individuality.

Chivian thinks it unlikely that there are any other microbial hangers-on in the mine since the methods he used have successfully isolated DNA from other microbes in the past. About 0.1% of sequences he identified did not fit into the main genome but Chivian thinks that these came from contaminating species, most of which had been picked up in the lab and others of which came from rocks higher up in the mine. If there are other species sharing the same space as *D.audaxviator*, they make up at most about 0.035% of the bacteria in the water and would be outnumbered by the dominant species by at a factor of 5,000.

D.audaxviator's genes revealed its abilities as well as its identity, and these support the idea that it is the sole occupant of the mine's depths. The bacterium has all the metabolic abilities required for an independent existence. It gets all the nutrients it needs from the minerals in the surrounding rocks, using sulphate ions in place of oxygen and getting nitrogen and carbon from ammonia and carbon dioxide dissolved in the surrounding water. It can even get an extra boost of nutrients by recycling dead cells of its own kind. It can form hard coats called endospores to protect it from dehydration or toxins, it has genes that allow it to sense nutrients, and it can create the components of flagella - whip-like tails that allow bacteria to move around. In fact, one of the only talents it lacks in comparison to other bacteria is the ability to deal with the dangerous chemical by-products of oxygen metabolism. But then again, it hardly needs such a skill in an environment devoid of the usually essential gas.

Many extremophiles survive by slowing their reproductive abilities, taking hundreds or thousands of years to divide, and paring down their genomes to just the bare necessities. But not *D.audaxviator* – it retains a genome size that is comparable to other free-living bacteria. Many of these abilities are the result of proteins that it has freely pilfered proteins from another kingdom of micro-organisms that is rife with extremists – the Archaea. These genetic loans, taken out before it moved into the gold mine, are the key to its success in this most inhospitable of environments.

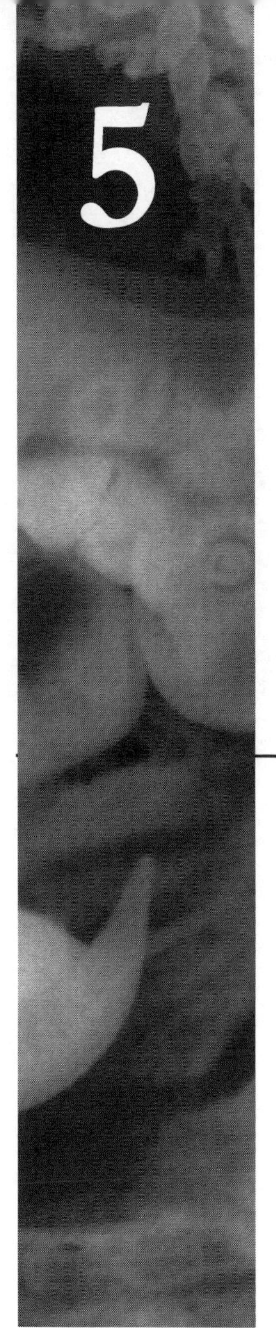

Being human

Stamina in a pill,
the genetics of commitment,
when fat cells become fatter and
the benefits of inbreeding

Ninety years of immunity

They say that memory declines as age marches on, but that only applies to brains – the immune system has a very different sort of memory and it stays fresh till the end of life. To this day, people who survived a global flu pandemic in 1918 carry antibodies that can remember and neutralise the same murderous strain of virus. The 1918 influenza virus was the most devastating infection of recent history and killed anywhere from 20 to 100 million people in the space of two years. Ironically, it seems that the virus killed *via* the immune system of those infected. It caused immune cells to unleash a torrent of signalling chemicals – cytokines – that recruited other immune cells to the fray. These too started signalling and caused a "cytokine storm" that raged out of control. This probably explains why the 1918 flu took such a heavy and unexpected toll on the young and healthy. Their strong immune systems would have done them little good against a virus that kills by causing those same systems to overreact.

Nonetheless, many children managed to fight off the infection and they are still alive to tell the tale. Xiaocong Yu from the Vanderbilt University Medical Center and Tshidi Tsibane from the Mount Sinai School of Medicine managed to track down 32 of these lucky survivors. Today, the youngest of the group is 91 and the oldest have seen their first century come and go. During the time of the pandemic, they could all remember one of their family being sick, making it likely that they themselves had been directly exposed to the usually lethal infection. And even though 90 years have passed, every single one of these people is still immune to the virus.[51] To this day, their blood samples can neutralise it.

Yu and Tsibane took the opportunity to examine the survivors' mighty antibodies. Using blood cells from seven survivors, the team managed to culture B-cells (a type of white blood cell) that secreted antibodies against the 1918 flu. They purified five of these antibodies, which had a particularly strong affinity for the 1918 strain and to a lesser extent, for a strain from 1930. Against later strains of flu from 1943, 1947, 1977 and 1999, these antibodies provided little defence. The antibodies recognise a protein on the surface of the virus called haemagglutinin. The 1918 and 1930 versions are structurally very similar but those from the later strains mutated to enough of an extent that antibodies fine-tuned to older viruses no longer worked.

The antibodies weren't just effective in test-tubes or Petri dishes – Yu and Tsibane proved their worth in living animals. They exposed mice to the reconstructed 1918 virus, a treatment that would usually kill them all. A day later, they injected them with the antibodies taken from the human

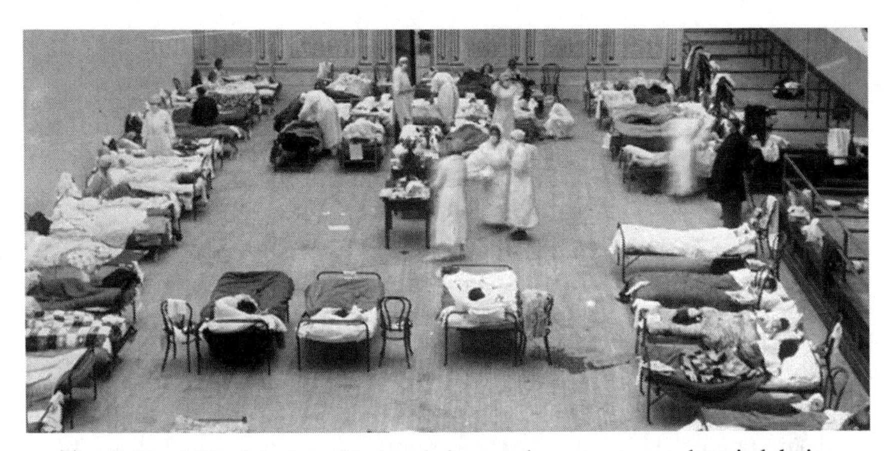

The Oakland Municipal Auditorium being used as a temporary hospital during the 1918 flu pandemic

survivors. Sure enough, compared to mice treated with unspecific antibodies, these animals had half the amount of virus in their lungs and lost several times less weight. And best of all, at high doses, all of them survived. The results indicate that B-cells that respond to infections endure for the lifetime of their host, even if that lifetime spans the best part of a century. The fact that these antibodies reacted strongly to the 1918 flu and not to later generations suggests they were the same defences that were stimulated some 90 years ago. However, Yu and Tsibane say it is likely that infections with related viruses in the intervening years helped to sustain the population of targeted B-cells. Just as they defended the survivors in 1918 and the mice in this experiment, these antibodies could defend humans in the future, should another similar strain of flu rear its head. And one of the five antibodies may even have a broader remit, for unlike its cousins it also managed to neutralise the 1977 virus. The antibody's structure could be broad enough to stick to a variety of different haemagglutinin proteins, or the virus may have effectively gone retro and 'recycled' an older structure that this particular antibody could recognise. Either way, its secrets are worth uncovering, for they could suggest strategies for granting immunity against a wider range of flu strains.

It is worth noting that these experiments were only possible because another group of scientists recreated the 1918 strain of flu in 2005. They used samples from a patient buried in Alaskan permafrost to decipher the virus's genome and structure. Resurrecting one of the most infamous viruses in history was a bold and (understandably) controversial move and, even though results like this are very interesting, it remains to be seen if it was a justified one.

Enhancing our thumbs

In September 2008, a group of US geneticists unveiled a picture of a mouse embryo with a small patch of blue in its paw, and rarely has a patch of blue caused so much excitement. The team had inserted a fragment of human DNA into the mouse's cells. The fragment contained an 'enhancer' element, a short span of DNA that switches other genes on and off. In this case, they put the enhancer in control of a gene whose activity creates a blue chemical. This particular enhancer was called HACNS1. Throughout the course of animal evolution, its sequence has gone relatively unchanged in almost all back-boned animals, but it has evolved rapidly in the human genome since we split away from chimpanzees. So the blue patch in the mouse's paw represented a site where this rapidly-evolved, human-specific piece of DNA was switching other genes on. Figured out why people were so excited yet? It was in the bit that would eventually become a thumb.

HACNS1 controls genes but it is not one itself. It is part of the large majority of our genome known as 'non-coding DNA'. A small proportion of our DNA is a code that tells our cells how to build its workforce of proteins, but the majority is never translated in this way. Much of this 'non-coding DNA' is functionless junk, but some types are very important indeed. The enhancers are one such group. They are stretches of DNA that control the activity of genes, which can often lie some distance away from the enhancer. When 'activator' proteins stick to the enhancers, the target gene is switched on. Change the sequence of these enhancers and you can change which genes they control, when they do so and where they do so. It is one way for evolution to exact big changes in a creature's body without having to add much in the way of genetic innovation. By changing enhancers, it can simply redeploy the existing squad of genes in new and interesting ways. The results can be dramatic, much like changing a sports manager can have a greater impact on a team's performance than switching out individual players.

There is evidence that these sorts of changes have indeed happened. Many human non-coding sequences show signs of incredibly rapid evolution ever since our evolutionary path diverged from that of chimps some six million years ago. Shyam Prabhakar from the Lawrence Berkeley National Laboratory singled out HACNS1 for attention because it is the most rapidly evolving sequence of its kind.[52] Only 16 differences separate HACNS1's sequence from that of its chimp counterpart, but that is about four times as many as you would expect if the sequence had just been drifting aimlessly while picking up new mutations. These rapid changes are a clear sign of "positive selection", where new mutations bestow such advantages on their bearers that they spread rapidly throughout the population.

So what does HACNS1 do? To find out, Prabhakar (together with a large team of geneticists) placed the human, chimp and macaque versions in a strain of mice. They set things up so that the sequence had control over a gene that creates a blue chemical when it is active. At the time, no one knew if HACNS1 was an enhancer; that only became apparent when they saw patches of blue in the young mouse embryos. Embryos that were loaded with the human version had strong blue stains in their developing limbs, eyes, ears and pharyngeal arches (structures that will eventually become the muscles, bones and organs of the mouth and throat). As the embryo develops, HACNS1 is clearly acting as a gene enhancer in these body parts. In comparison, the chimp and macaque versions drove very little gene activity at these sites, particularly at the limbs, where many embryos had no signs of blue colour at all. When the team looked at older embryos, they saw that the enhancer was still activating genes in the young mouse's limbs. They saw the telltale blue stain in the shoulder, wrist and thumb of the front limbs and to a lesser extent in the big toe, ankle and hip of the hind pair. Once again, the chimp and macaque enhancers were far less enhancing, and only drove a smattering of gene activity in the shoulder area.

The results are preliminary but they are exciting. They suggest that changes in HACNS1 *may* have contributed to the uniquely human aspects of our thumbs, wrists, ankles and feet. Our long and fully opposable thumbs allow us to grip and manipulate objects with great precision while our inflexible feet and short toes give us the stability that life on two feet demands. There is no doubt that these physical innovations have played a key part in our success as a species, and perhaps HACNS1 can take some credit for that. Prabhakar's team found further proof of the uniqueness of human HACNS1 with a deceptively simple game of genetic swapsies. They added all 16 human-specific changes to the chimpanzee version, and also stripped out all 16 changes from the human sequence to revert it back to the chimp one. In the mouse embryos, the "humanised" chimp enhancer produced the same pattern of blue as the natural human version did, while the "reverted" human enhancer yielded the same pattern as the untouched chimp and macaque ones. These 16 genetic tweaks have raised HACNS1's profile, turning it from a bit-part actor into one of the major stars in the drama of development. It is not entirely clear how this has happened, but a quick analysis showed that the 16 changes have probably altered the way that the enhancer interacts with its activator proteins. The proteins attach to the enhancer through specific docking sites that are dictated by its sequence.

As this sequence changed in 16 ways, docking sites were added for some proteins and lost for others.

The next part of the quest will be to find which genes are enhanced by the enhancer. The closest two are CENTG2, whose role in limb development has never been looked at, and GBX2, which is activated in embryonic limbs. Nor is the role of HACNS1 confined to limbs – it is also activated in eyes, ears and the precursors of mouths. And perhaps there are other organs that are affected in humans but do not show up in the mouse embryos. These are all questions for another time. For now, the results can be taken as yet further evidence that many of evolution's big breakthroughs, from eyes to language, are the result of genetic tinkering rather than novelty.

The teenage obes-o-stat

Fat people have an abundance of fat tissue, so the natural assumption is that fat people have more fat cells, or 'adipocytes'. But that is only part of the story – it turns out that overweight and obese people not only have a surplus of fat cells, they have larger ones too. The idea of these 'fatter fat cells' has been around since the 1970s. But their importance has been dramatically highlighted by a new study, which shows that the number of fat cells in both thin and obese people is more or less set during childhood and adolescence. During adulthood, about 8% of fat cells die every year only to be replaced by new ones. As a result, adults, even those who lose masses of weight, have a constant number of fat cells. Instead, it is changes in the *volume* of fat cells that causes body weight to rise and fall. Kirsty Spalding from the Karolinska Institute in Sweden, together with a large team of international researchers, uncovered several lines of evidence to support these conclusions.[53] Her study is a fascinating mix of cell counting, stomach surgery, radioactive Cold War fallout and a rather surprising use for carbon-dating.

First, off she compared body fat measurements from over 800 people to the size of their fat cells, as viewed under a microscope. She found that people with more body fat also had larger fat cells and this held true for fat deposited both under the skin (subcutaneous fat) and around the belly (visceral fat). However, body fat and fat cell volume weren't perfectly matched – the *number* of fat cells was also important. Spalding counted the total number of fat cells in 687 adults and combined these tallies with measurements for children and adolescents taken from previous studies. Together, the data showed that a person's pool of adipocytes is set during childhood and adolescence, increasing during these periods and levelling off during adulthood. Both lean and obese people showed the same pattern and adults in both categories had very little variation in their fat cell count. Even people who lost massive amounts of weight still had the same number of fat cells. Spalding studied the fatty tissue of 20 people before and after they went through bariatric surgery, a set of drastic procedures that shrink the stomach with staples and bands, or bypass it altogether. It is an extreme approach to weight loss, but it works, and the people in the study dropped an average of 18 Body Mass Index (BMI) points. Their fat cells had shrunk by about a third but they still had the same number, a year or so after the operation!

An unchanging number of fat cells means one of two things – the same cells persist throughout adulthood, or they are destroyed and replenished at the same rate. Spalding suspected the latter, and to prove it, she relied on the most improbable of sources – fallout from the Cold War. During the late 1950s, the world's superpowers busied themselves by testing their new nuclear arsenals, unleashing a large amount of radioactive isotopes which spread around the globe. These included ^{14}C, a form of carbon typically found at low background levels in the atmosphere. From 1955, atmospheric levels of ^{14}C shot up to heights unheard of for thousands of years, only to fall exponentially after the Test Ban Treaty of 1963. In the meantime, the extra ^{14}C was converted into carbon dioxide, taken in by plants, and made its way up the food chain. Some of it ended up in the bodies of people alive at the time, and if any of them were creating new fat cells, these would be laced with unusually high levels of ^{14}C. In this way, the levels of ^{14}C in any adipocyte acts as a time stamp, directly reflecting the amount in the air at the time it was created. The bomb tests had inadvertently given Spalding a way of carbon-dating fat. And she could do so very accurately, because of the massive year-on-year changes in atmospheric ^{14}C. She took fat tissue from 35 people who were either lean or obese and used them to create cultures of fat cells that were almost completely pure. That is important as contamination with other cell types that turn over at different rates would have messed up the results. She found that all the samples taken from people born before 1955 had levels of ^{14}C that were substantially higher than the low atmospheric levels before the bomb tests. People born after the Cold War showed the same pattern – their fat cells were often about 20 years or so younger than they themselves were. These cells were clearly produced during either adolescence or adulthood.

Armed with all of this data, Spalding managed to model the life and death of fat cells. She discovered that obese adults produce about twice as many new fat cells every year as lean ones. However, obese adults also have more fat cells anyway than lean ones, and the *proportion* of new cells added is the same in both groups. These similar birth rates are matched by similar death rates, so regardless of weight, adults replace about 8% of their fat cells every year and the total number stays the same. In early life, things are different. Compared to their lean peers, obese children add new fat cells at twice the rate, which is why they end up with a bigger complement in adulthood.

Spalding's stunning results help to explain why so many overweight and obese people find it very hard to lose weight. Having built up a large supply of fat cells that is constantly replenished throughout adult life, they are already at a disadvantage. Previous studies have shown that an overabundance of fat cells leads to a deficiency of leptin, a hormone that normally keeps appetites in check and boosts metabolism. With leptin in short supply, people eat more and their numerous fat cells become swollen with extra lipids. There is one caveat – the obese people included in the study were all obese from an early age. So it is unclear if the fat cells of people who gradually gain weight during adulthood will eventually reach some maximum size and trigger an increased production rate for new cells. It is a possibility, but it is unlikely to be a very important one because surprisingly few people become obese as adults. While 75% of obese children grow up to be obese adults, only 10% of those with a healthy weight do the same.

Spalding also suggests that her discoveries could be used to develop innovative treatments for obesity. Some cell types are subject to simple feedback loops that limit their number. Muscle cells, for example, secrete a molecule called myostatin that in sufficient quantities, blocks the production of new muscle cells. It is likely that fat cells operate under the watch of a similar cellular thermostat and the molecules involved, if we can identify them, would make enticing targets for anti-obesity drugs. Such developments are likely to be years away, and for now, the rising levels of obesity around the world pose a more immediate problem. Through this study, we could not have a clearer indication of the importance of childhood as a window for preventing obesity and the chronic diseases affected by it – cancer, heart disease, diabetes and more. The message is especially stark following the recent Foresight report in the UK, which estimated that if current trends are left unchecked, by 2050 a quarter of all the country's children under the age of 16 will be obese. The knowledge that their fat cell count will then be set for life makes the cost of inaction even higher. Fortunately, it seems that the UK Government is taking appropriate steps and recently pledged over a third of a billion pounds on a concerted strategy to tackle childhood obesity. The study does not just affect how we deal with obesity pharmaceutically and politically, but socially as well. The more we learn about obesity, the more ludicrous it becomes to blame the condition on a lack of discipline or shoddy willpower. It is clear from studies like these, and our ever-increasing understanding of the genetics of obesity, that being fat has a strong biological basis that can be very difficult to overcome.

Stamina drug

For the first time, scientists have developed drugs that mimic the effects of endurance exercise. With the aid of two chemicals, Vihang Narkar, Ronald Evans and colleagues from the Salk Institute managed to turn regular lab rodents into furry Paula Radcliffes – mighty mice that were capable of running further and for longer than their peers. Using drugs to boost performance is not a new development. Steroids can help body-builders to build their bodies, while giving athletes an extra burst of speed. But this is the first time that scientists have managed to develop chemicals that improve stamina, as opposed to strength or speed. One of these drugs only worked when taken in conjunction with exercise, but the other boosted endurance in inactive, couch-potato mice too.

The changes lay in their muscles. The body's skeletal muscles (the ones attached to bones) are made of two major types of fibre – the "fast-twitch" kind that burn sugar, and "slow-twitch" ones that prefer to burn fat. The bulkier fast-twitch varieties contract more quickly and powerfully, and provide short bursts of speed and strength; they are plentiful in the bodies of sprinters and body-builders. The slow-twitch versions are more resistant to fatigue and can carry on working for hours rather than minutes; they are the province of marathon runners and mountain climbers. Endurance training triggers a suite of genetic changes that converts the fast-twitch fibres into the slow-twitch ones, imbuing muscle with greater staying power. Narkar's two chemicals emulate the effects of exercise by tapping into the same genetic pathways for reprogramming muscle.[54]

The first drug carries the uncharismatic name of GW1516 and activates a gene called PPARd. From the start, the gene was a promising target. It controls the metabolism of skeletal muscles, and other groups have managed to double the endurance of laboratory mice by permanently switching it on. So it must have come as a bit of a shock to Narkar's team that feeding mice with GW1516 did not improve their endurance at all. Things changed when exercise was thrown into the mix. With a combination of GW1516 and four weeks of daily exercise, mice managed to run for about 70% further and longer than mice that exercised without the drug. By itself, the training improved their stamina by an hour or so, but with the performance-enhancer, they carried on for almost an extra two. The combination of drug and exercise also substantially increased the proportion of slow-twitch fibres in their muscles and upped the activity of genes involved in burning fats. The combination of exercise and GW1516 did not just add their individual benefits together – it was a case of the whole being more than the sum of the parts. The combo switched on a unique "signature" of endurance genes involved in fat metabolism and muscle remodelling, many of which weren't affected by either GW1516 or exercise alone.

The second drug, called AICAR, is even more exciting, for it managed to increase the endurance of mice without a need for exercise. When mice were given the drug, even the inactive ones managed to run 44% farther and 23% longer than their peers. AICAR works by activating a gene called AMPK, which is also involved in muscle metabolism and switched on by exercise. In addition to AMPK, Narkar saw that AICAR switched on a set of 32 genes in the muscles of mice, the majority of which are also controlled by PPARd. In fact, PPARd and AMPK form a close molecular alliance, in which AMPK dramatically boosts the effects that PPARd has on its own target genes. This explains why GW1615 only succeeded in improving the endurance of mice in conjunction with exercise. GW1615 switches on PPARd but to little effect without the extra boost given by AMPK. Add exercise into the mix, and AMPK is switched on allowing the alliance to work in tandem. This collaboration also explains why AICAR improved endurance without the need for extra exercise. Narkar's idea is that AMPK is so central to the genetic events triggered by exercise that activating it directly with AICAR bypasses the need for any exercise.

That is not to say that exercise is irrelevant – AICAR may have boosted endurance on its own, but not to quite the same extent that exercise normally does. Nonetheless, by chemically mimicking some of the benefits of exercise, the team's hope is that the drugs will find a use in treating metabolic diseases such as obesity, or warding against muscle wasting and frailty. After all, keeping active has a wide range of benefits that go well beyond its effects on endurance. There is even some evidence to suggest that Narkar's drugs may do more good than simply improving stamina. In some of the mice, they triggered fat loss alongside increased endurance. PPARd controls hormone levels and heart muscle as well as skeletal muscle, while AMPK's ability to lower blood sugar levels may be relevant to diabetics. The potential benefits (and risks) of manipulating these genes will not be fully known until Narkar checks their effects in these other tissues.

For now, the main risk has to be the potential for cheating in athletics tournaments. Narkar's team have not confirmed if the drugs would have the same effects in humans, but there is evidence that GW1615 at least is active in the human body. It is not clear if this sort of pharmaceutical stimulation could ever top the type of benefits that hard athletic training can bring, or even if they would have any effects in people who have already pushed their bodies to the limits of endurance. Even so, the researchers have already spoken to the World Anti-Doping Agency about developing a test that detects the use of PPARd-boosting drugs.

Skin to stem to nerve

Potential is a sad thing to lose. Have you ever thought that it would be great to return to your childhood, when your options seemed limitless and life hadn't taken you down increasingly narrow corridors of possibility? Wouldn't it be great to rewind the clock and have the choice to start over? While that is still the stuff of dreams, for some cells in your body it may soon be reality. In November 2007, two groups of scientists found a way of turning skin cells from human adults back into the stem cells of embryos; shortly after, another group used these reprogrammed cells to produce neurons.

Embryonic stem cells are the embodiment of potential. Armed with a trait called 'pluripotency', they can give rise to every single type of cell and tissue in the body, renewing themselves indefinitely while their daughters take up the specialised guises of nerves, muscles, blood and more. For years, stem cells have been touted as the Holy Grail of modern medicine. Within their membranes lies the potential to understand how we develop, test new drugs and most importantly, provide replacement cells to treat Alzheimer's, Parkinson's, spinal cord injuries, diabetes, stroke and more. To fulfil this potential, we need to find a way to produce stem cells with the same genetic material as the patient being treated. There are two ways to achieve this personal touch. The first involves transplanting a nucleus (and the DNA inside) from one person's cell into an empty egg. Egg turns into embryo, and embryo provides embryonic stem cells. But harvesting these cells destroys the embryo, which steers the process into an ethical quagmire.

The second method is less controversial – we could reprogram adult cells into their original stem-like state, rather like erasing a person's CV and school record and have them start again as a fresh-faced child. As stem cells spawn off new lineages, the DNA of these daughters picks up molecular tags that dictate how, when and where their genes are to be read. These 'epigenetic' changes commit the cells to increasingly specialised fates and

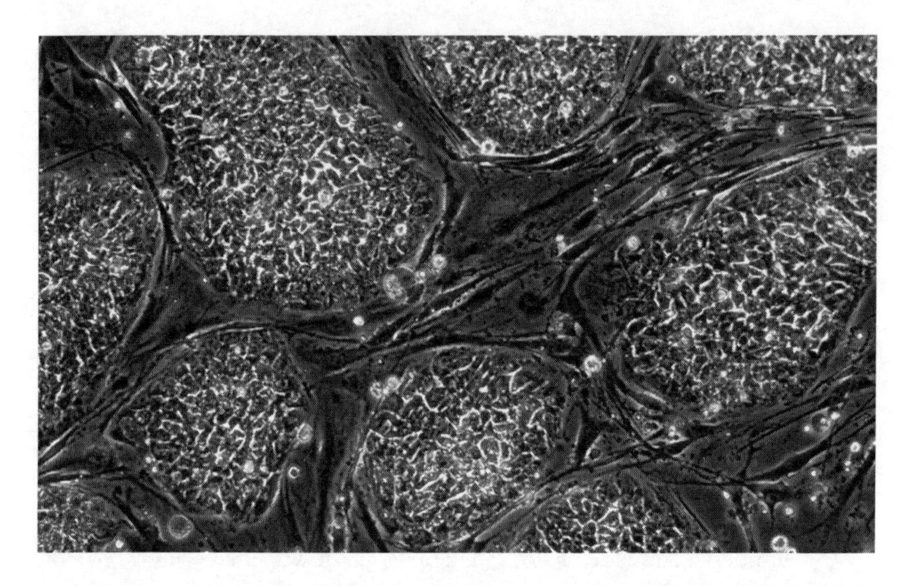

ensures that development is an ordered, timely process. In theory, these changes should be reversible with the right combination of chemicals.

Shinya Yamanaka and Kazutoshi Takahashi from Kyoto University were the first to prove that this could be done in mice. Using just four proteins – Oct4, Sox2, c-Myc and Klf4 – the duo managed to turn cells from connective tissue taken from their tails into stem cells or, as they called them, 'induced pluripotent stem (iPS)' cells. That experiment fired the starting pistol in a race to duplicate the success in human cells and two groups shot past the finishing line in a photo-finish tie. Yamanaka and Takahashi were in the driving seat of the first team and they showed that the same four proteins could turn adult human cells into iPS cells.[55] They successfully reprogrammed skin cells from the face of a 36-year-old woman and connective tissue from a 69-year-old man, turning them into iPS cells. These were remarkably similar to true embryonic stem cells; they had the same shapes, rates of growth, portfolios of active genes and coats of protein. And just like embryonic stem cells, they had high levels of telomerase, an enzyme that effectively grants them immortality by repairing protective structures at the ends of their DNA. As a final proof of their 'stemness', Takahashi and co. successfully coaxed the induced stem cells into producing neurons and heart muscle cells. When they were transplanted into the sides of mice with weakened immune systems, they formed tumours called teratomas (literally 'monster tumour') containing a Frankenstein-like mash-up of different cell types including cartilage, gut lining, muscle, nerves and keratin.

Using c-Myc is a bit of a problem though, as mutant forms of this gene are linked to several human cancers and the protein can sometimes cause human embryonic stem cells to die. With this in mind, a second group led by Junying Yu and James Thomson from the Genome Center of Wisconsin searched for a different combination of reprogramming proteins.[56] They whittled their way through 14 candidates and through a process of elimination hit upon a slightly different quartet – Oct4 and Sox2, as used by the Japanese group, and two newcomers, Nanog and Lin28. In similar experiments, they used these proteins to reprogram skin cells from a foetus and from the foreskin of a baby boy. Again, the induced stem cells were superficially and genetically similar to true embryonic stem cells, were rife with telomerase, and could produce a wide variety of cell types.

So far, so promising, but it is not quite time to hang up the lab coats and start distributing panaceas. The two papers were both breakthroughs in the field, but neither has come up a method that's feasible for actual clinics. The biggest drawback of the studies is that both used retroviruses to cut and paste the quartet of genes into the reprogrammed cells. This is quite a brutish method. The viruses inevitably insert their load in places that disrupt other genes, which could eventually lead to cancer in tissues grown from the resulting cells. Indeed, Yamanaka found that one in five of the mice grown from his induced mouse stem cells eventually developed cancer. That is obviously unacceptable and stem cell researchers need to find new methods. They could either develop carriers that shuttle the new genes and proteins into the cells without disturbing their DNA, or create small molecules that activate the existing genes without the need for introducing new copies.

There is also a small but significant question over the identity of the induced stem cells. They may be very similar to their embryonic counterparts, but they are not identical. The Japanese group compared 32,000 genes in the two groups of cells and found large differences in activity in over a thousand of these (just under 4%). It will be important to discover the precise nature of these differences. We will also need a better understanding of the method itself. In a case of the technology preceding the science, it is not entirely clear how these two sets of four proteins restore an adult cell's lost potential. So far, we know that Oct4 and Sox2 work together to activate genes that maintain 'stemness', while suppressing genes that chain cells to specific fates. Non-stem cells build up a series of barricades that stops these partners from affecting their target genes, and Takahasi speculates that c-Myc and Klf4 alter the way DNA is packaged to give Oct4 and Sox2 easier access to their targets.

In the meantime, other research groups have taken these breakthroughs and run with them. Within less than a year, John Dimos and Kit Rodolfa from the Harvard Stem Cell Institute had surpassed two further milestones.[57] As before, they transformed adult skin cells into embryonic stem cells but with two important differences. Firstly, the cells came not from a young, healthy individual, but from an 82-year old woman with chronic disease – amyotrophic lateral sclerosis (ALS), the same condition that has paralysed Stephen Hawking. Even after a lifetime of chronic disease, the adult cells could still be reverted to a stem-like state. It is more of an expansion on a theme than an isolated breakthrough but it is nonetheless important. Many of the diseases behind the stem cell hype are chronic conditions that affect the elderly – Alzheimer's, Parkinson's and the like. The ultimate goal of this research is to use the cells of patients with these conditions to generate personalised embryonic stem cells. These can then be used to grow cells-of-choice to replace those lost through disease, without fear of reprisals from the patient's immune system. The new study shows that the first of these steps is possible, even for very old people suffering from advanced disease. More importantly, Dimos and Rodolfa managed the second step too; they converted the reprogrammed stem cells into motor neurons, the same cells that are affected by the donor's disease. This double-whammy of transformations is exactly the type of technique we need in order to realise the potential of stem cell therapies.

The woman in question is named only as A29. Dimos and Rodolfa took skin cells from her arm and exposed them to the same quartet of transformative genes used by the Takahashi's group. After two weeks, they found what they were looking for – small colonies of iPS cells. For their next trick, the team transformed the iPS cells even further; they used small molecules to trigger two developmental programmes that coaxed the cells into becoming a variety of neurons. Dimos and Rodolfa confirmed the nature of their new cells by checking their shape under the microscope and staining them with a dye designed to only stick to neurons. About 20% of the cells had activated genes that singled them out as motor neurons, and mature ones at that.

The same problems still apply – how are the iPS cells different from true embryonic stem cells, and will the technique increase the risk of cancer? So there is still a long way to go but for the moment, the study is also a godsend for ALS research. The disease is caused by the gradual death of the body's motor neurons, cells that carry signals from the central nervous system to the muscles. Without these signals, the muscles of ALS sufferers gradually weaken and waste away. Voluntary movements become impossible and in most cases, people lose the ability to swallow or breathe. It is a horrible disease that usually kills within three to five years. So far, progress in understanding the disease has been relatively slow, mainly because it has been nigh impossible to obtain a decent supply of living motor neurons affected by the condition. But Dimos and Rodolfa's work changes all of that. Now, researchers can culture large colonies of motor neurons and other brain cells that carry genetic defects associated with ALS. That gives them free reign to investigate the genetic defects that underlie the disorder, the environmental conditions that interact with these genes, and the way the affected neurons interact with other types of cell. It also provides them with neurons to use for screening and testing potential drugs. It is a starting pistol, and a loud one at that.

As far as stem cell research is concerned, it seems that reprogramming has taken the lead over nucleus transfer as a technique for producing stem cells. It certainly avoids the moral dilemma posed by its rival method by avoiding the need for embryos and donated eggs. Nonetheless, it would be prudent to remember that these new successes depended on knowledge that we gleaned through research on actual embryonic stem cells. This line of work will also be necessary to help us answer the many unsolved questions posed above. It is premature to dismiss nuclear transfer as an unnecessary technique and certainly rich to suggest, as George Bush has done, that trying to *block* this research was actually responsible for driving scientists down the more morally acceptable path of reprogramming.

Fat, fatter, fattest

Many measures to curb the obesity epidemic are aimed at young children. It is a sensible strategy – we know that overweight children have a good chance of becoming overweight adults. Family homes and schools have become critical arenas where the battle against the nation's growing waistlines is fought. But there is another equally important environment that can severely affect a person's chances of becoming overweight, but is more often overlooked – the womb. Overweight parents tend to raise overweight children but over the last few years, studies have confirmed that this tendency to transcend generations is not just the product of a shared home environment. Obesity-related genes are involved too, but even they are not the whole story. A mother's bodyweight in the period during and just before pregnancy has a large influence on the future weight of her children. For example, children born to mothers who have gone through drastic weight-loss surgery (where most of their stomach and intestines are bypassed) are half as likely to be obese themselves. On the other hand, mothers who put on weight between two pregnancies are more likely to have an obese second child. In this way, the obesity epidemic has the potential to trickle down through the generations, like a snowball rolling its way into an avalanche. Robert Waterland from the Baylor College of Medicine demonstrated how the snowball gains momentum by studying three generations of mice that have a genetic tendency to overeat. And by using a special diet that was high in folate and other nutrients, he found that he could stop the snowball's descent and spare future generations of mice from a heightened risk of obesity.

Mice carry a gene called *agouti* that causes their hair to produce a yellow pigment rather than a black one. Waterland's mice carry a mutant version of *agouti* called A^{vy}, which is switched on throughout their entire bodies, rather than just their hair follicles.[58] That gives the mice a mottled yellow colour, but as an unexpected side effect, the ever-present gene also blocks signals in the animals' brains that tell them that they are full. That makes the mice prone to overeating, and if they are given free access to an all-you-can-eat buffet, they invariably get fat. There are obvious parallels to modern humans, who have constant access to cheap, calorific food at massive portions after evolving to cope with conditions where food is often lacking. Waterland followed three generations of these gluttonous mice; at each stage, females with one copy of A^{vy} were mated with males without any. The result was a family, where grandmothers, mothers and daughters alike all carried the same version of *agouti,* which predisposed them to overeating. But despite these genetic similarities, and despite the fact that they were all raised on the same diet, successive generations of mice became progressively fatter. In the first generation, 45% of A^{vy} carriers weighed more than 50g (which is heavy for a mouse). By the second generation, 54% passed this mark and by the third, a whopping 72% had gone beyond it. In each generation, the mother's weight predicted the weight of her offspring, so that heavier mums begat heavier pups.

These results are unique. Other studies have used artificially manipulated diets to examine the link between obesity in mothers and children, but Waterland's work is the first to show that natural variations in obesity can cascade down the generations. If the same process applies to humans, and there is evidence that it does, it is likely that the increase in obesity rates among women of a childbearing age could help to explain the enormous rises in average bodyweight over the past few decades. Now remember that the researchers took steps to ensure that the A^{vy} gene was the same in all three generations of mice. This means that the process is not a genetic one. Instead, the changes are epigenetic, a term which means that a gene itself stays the same, but the way it behaves is altered. Genes can be switched on or off and turned up or down by adding chemical groups to their DNA sequence. For example, simple 'methyl groups' consisting of one carbon atom and three hydrogens can serve to silence genes. These add-ons are like

Post-It notes stuck to a book that tell you how to read it, even though the underlying text is the same. Many of these epigenetic changes remain when cells divide and they can persist across generations (copy the book and the Post-its are copied with it). This process provides a route for environmental factors – in this case, a mother's body fat – to leave a legacy that is passed down the family line.

Waterland confirmed that epigenetics was involved by showing that a cocktail of nutrients, including folate, vitamin B12, betaine and choline, could curtail the generational fattening in his mice. These nutrients provide a rich source of methyl groups, which serve to silence various genes throughout the rodents' genomes. A^{vy} was certainly one of these. But Waterland thinks that other genes are shut down too, including some that are involved in the development of the hypothalamus, a part of the brain that controls appetite. The exact details of this process, and which genes are affected, are still unknown. But the end-results are clear. With this special diet, the first and second generation mice had similar proportion of obese individuals (45% and 52% respectively) as their peers, but in the third generation, a mere 44% were obese compared to 72% in the other mice that did not get any supplements. And among the supplemented families, the weight of mothers did not influence the weight of their offspring and successive generations did not get fatter and fatter.

The field of epigenetics is one of the hottest in modern biology. We now know that problems with epigenetic marking play a big role in the development of cancer, and it is likely to change the way we see other diseases too. According to Waterland, we know "virtually nothing" about the epigenetic control of genes that are involved in eating or hunger. Given the increasing rates of obesity across the globe, filling in this void of ignorance is of crucial importance.

Of voles, men and love-rats

Love is all around us and love is in the air, but is it also in our DNA? New evidence suggests that variation in a gene called AVPR1A has a small but evident influence on the strength of a relationship, the likelihood of tying the knot and the risk of divorce. It is news for humans, but it is well-known that the gene's animals counterpart affects the bonds between pairs of voles. The story really starts with these small rodents and it is them that I now turn to.

Voles make unexpectedly good animals to study if you are interested in the genetic basis of commitment, because closely related species have very different mating strategies. The prairie vole is (mostly) monogamous; males and females mate for life and look after their pups with great care. On the other hand, the closely related meadow and montane voles have more of the love-rat about them – males have many mates, flit between them and take no responsibility for the care of their many young. Behind these different behaviours lie is a hormone called vasopressin and its partner molecule – the vasopressin receptor V1aR. Vasopressin is a neurotransmitter – a signalling molecule of the brain – and it transmits signals by attaching to V1aR, like a key fitting into a lock. Alter the balance of these molecules and you can change the voles' behaviour. For example, give extra vasopressin to a prairie vole and it will develop a stronger bond with its partner, but block the receptor and you break the bond. It even works in the promiscuous meadow voles. Prairie voles have much higher levels of the vasopressin receptor than meadow voles do and in 2004, Miranda Lim loaded the forebrains of meadow vole males with extra copies of V1aR to the same levels of their prairie cousins. The males were paired with a female and when they had to choose between her and a newcomer, those that packed extra V1aR spent more of their time with their familiar partners. Those that were injected with a random gene, unrelated to monogamy, stuck to their old promiscuous ways.

The sequence of the V1aR gene is very much the same in the three vole species but there are differences in the area just ahead of the gene, which affects how it is deployed. In this regulatory area, the prairie vole has a short stretch of DNA that its promiscuous cousins lack. Altering this sequence affects not only the distribution of V1aR in the animal's brain, but also its fidelity towards its partner. Humans have our own version of the vasopressin receptor, with its very own unmemorable acronym – AVPR1A. Like its vole counterpart, it is preceded by an important stretch of DNA that is rife with repetitive sequences. These are known as "repeat polymorphisms"; they are short genetic leitmotifs that vary in number from person to person. According to earlier research, these variations in this sequence can affect human behaviour and are linked to altruistic tendencies, the risk of autism and the age at which people first have sex.

We can now add the strength of relationships to that list. Hasse Walum at the Karolinska Institute compared the sequence of repeat polymorphisms in AVPR1A among 552 pairs of Swedish twins, all of whom were either married or living with a partner.[59] He also asked them a series of 13 questions designed to find out about various aspects of their relationships such as how much time they spent together, how often they kissed or whether they had ever considered divorce. Based on their answers, each person was given a score that indicated the quality of their relationship, with a maximum of 66. Walum found a link between this score and a version of one particular repeat polymorphism. Men who carried this sequence, known as allele 334, had lower-than-average scores and the more copies they had, the lower their rating. It is worth noting that the differences were tiny – men with no copies of allele 334 had an average score of 48 out of 66, while those with two had an average score of 45.5. Nonetheless, statistical tests suggested that the differences were unlikely to have arisen by chance. Carriers of allele 334 also tended to suffer from more marital problems than those without it. About a third of men (34%) with two copies had faced rocky marriages or impending divorce within the last year, while only half as many (15-16%) of those with one or no copies had been through such problems. People who carried allele 334 were also slightly less likely to have tied the knot. About 83% of men with one or no copies had married their partners (the rest were co-habiting) while only 68% of those with two copies had taken that step. The fact that it only really affects the behaviour of men is consistent with vasopressin's role in voles. And while women are not directly influenced by it, they are certainly indirectly affected. Those who were partnered with allele 334 carriers were prone to rating their relationships slightly more harshly than those whose partners did not carry it.

Based on these results, Walum suggests that AVPR1A does indeed affect the bonds that we form with our partners although to a much weaker extent than its counterpart does in voles. Allele 334 in particular has a negative influence on our relationships. The effects may be small but bear in mind that everyone in this study already showed a high level of commitment by marrying or moving in with their partner. Widen the sample to include chronic singletons and you might see much stronger effects, especially since about 40% of men carry at least one copy of allele 334. Even so, it is clear that the researchers are being very cautious about their findings. The word "tentatively" appears no less than three times in the closing paragraphs of their paper. Indeed, they specifically write that "this study clearly does not mean that this polymorphism may serve as a predictor of human pair-bonding behaviour on the individual level". You cannot sequence someone's AVPR1A gene and tell if they are going to be a suitable partner.

This study made a big splash in the mainstream press, where the majority of pieces proclaimed the discovery of a "divorce gene", a "monogamy gene" and yes, the "love rat gene", among other equally inane descriptions. They are unhelpful terms, for they prolong the misconception that single genes can account for complicated behaviours, and that they do so in a deterministic way. The BBC even reported that uncommitted men could blame their genes (accompanied by an amusing picture of George Clooney). The study, of course, says nothing of the sort. It *does* however confirm that researching the love lives of voles is a good idea – the lessons we have learned about the effect of vasopressin in their tiny brains are also relevant to humans. It means that looking for other players that affect relationship-building in voles is a worthwhile quest. Vasopressin is far the only molecule involved in forming relationships, even in voles and there is still much we do not know about the other players involved. Monogamy and promiscuity themselves are not clear-cut distinctions and there has been recent evidence that even the prairie vole is not the faithful saint it is made out to be...

Too much reception

There can be few events more devastating for a parent than cot death – the sudden and unexpected death of a baby. Cot death is more formally known as Sudden Infant Death Syndrome (SIDS) and it is an apt title, for affected babies often seem outwardly healthy and show no signs of suffering. Studies have suggested that things like passive smoke and breastfeeding can affect the risk of SIDS but the underlying biology behind the syndrome is a mystery, as is the cause of death in most cases. But it is a mystery that is slowly being solved. The latest and most intriguing clue comes from Enrica Audero from the European Molecular Biology Laboratory, working together with European researchers and a special strain of mice. Audero has shown that altering the balance of the signalling molecule serotonin in the brainstems of mice can lead to sudden demise, in a way that resembles the unexpected death of SIDS babies.[60] The mice spontaneously go through "crises" where their basic body functions like temperature control and their heartbeat go haywire.

Audero's work builds on research published by an American group two years ago, which first suggested that SIDS is the result of faults in the way our brains reacts to serotonin. This signalling chemical helps to control the core functions of our body that lie outside the realm of conscious thought. It is the serotonin system that lords over our heartbeats, breathing, sweating and shivering, while our brains are busy reading books or solving Sudokus. Through a series of post-mortem exams, David Paterson showed that SIDS babies have more serotonin-releasing neurons, but a lower density of serotonin receptors – protein docks that the molecule sticks to. It was a start, and Audero capitalised on it by showing for the first time how an altered serotonin network could actually lead to sudden death.

Her team developed a strain of mice with abnormally high levels of Htr1a – a serotonin receptor. With ten times more Htr1a than their littermates, you might expect these mice to be extremely responsive to serotonin, but in fact, the opposite was true. Serotonin is one of many chemicals in our body that regulates itself. It is secreted by a cluster of nerve cells within the brainstem called the raphe nuclei, which are also loaded with Htr1a receptors. The docking of serotonin into these receptors stops the raphe nuclei from pumping out more serotonin. It is a signal that limits its own transmission and this self-regulation means that a surplus of Htr1a actually resulted in much weaker responses to serotonin signalling.

The big surprise was that the majority of these altered mice shared the sad fate of human SIDS infants, with most dying before reaching three months of age. Their heart rate and breathing were mostly normal but would drop abruptly and frequently, something which never happened with normal mice. Every time it happened, the mice took days to recover and each time, there was a one in three chance that they would die from an irreversible fall in both heart rate and body temperature. SIDS babies too, tend to show falling heart rates and breathing patterns. And like SIDS babies, the altered mice were most vulnerable during a critical time window. If the extra Htr1a copies were switched on after 40 days, 90% of the mice died but only 30% did if the team waited until the 60-day mark to flip the switch. A twenty-day window separated an almost certain chance of death, with a fairly good chance of survival. Audero confirmed that it was the extra serotonin receptors that were the problem – her mice were engineered so that the extra copies could be removed with an antibiotic called doxycycline and this treatment restored their normal life spans.

She was not able to directly identify the cause of these crises, but she believes that environmental stress may be the key. The extra receptors disrupt lines of communication between serotonin neurons and those in the spine that are connected to major organs like the heart and lungs. These networks control the body's responses to the outside world and if they do not work properly, the mice cannot right themselves in the face of environmental changes. For example, when a normal mouse gets cold, its serotonin neurons would tell its brown fatty tissue to generate more heat, by switching on a gene called UCP1. But this process completely fails in mice with an abundance of Htr1a receptors – when they get cold, they stay cold, and a fall in body temperature is just one of the symptoms of a crisis event.

The results are a bit of a surprise. After all, mice can survive quite happily if they are born without any serotonin neurons at all, even though their behaviour may be different. And Audero is not suggesting that human SIDS babies have extra serotonin receptors; you'll remember from Paterson's study that they actually have a *lower* density of these. But the results do suggest that an imbalance in serotonin signalling can be life-threatening. Even without any overt outside shock, it is enough to trigger the sort of sudden failure in core body functions that is typical of SIDS. The means may be different in humans, but the ends are the same. Now, at least, we have a useful animal model for studying the condition and answering some key questions – are crises more likely to happen during sleep (when most human SIDS cases take place), are there signs that can predict the onset of a crisis, and are there ways to predict which babies have a high risk of the condition?

Women vs. girls

You could argue that life at its most basic is all about cheating death and having enough sex to pass on your genes to the next generation, as many times as possible. From this dispassionate viewpoint, human reproduction is very perplexing for our reproductive potential has an early expiry date. At an average age of 38, women start becoming rapidly less fertile only to permanently lose the ability to have children some 10 years later during menopause. From an evolutionary point

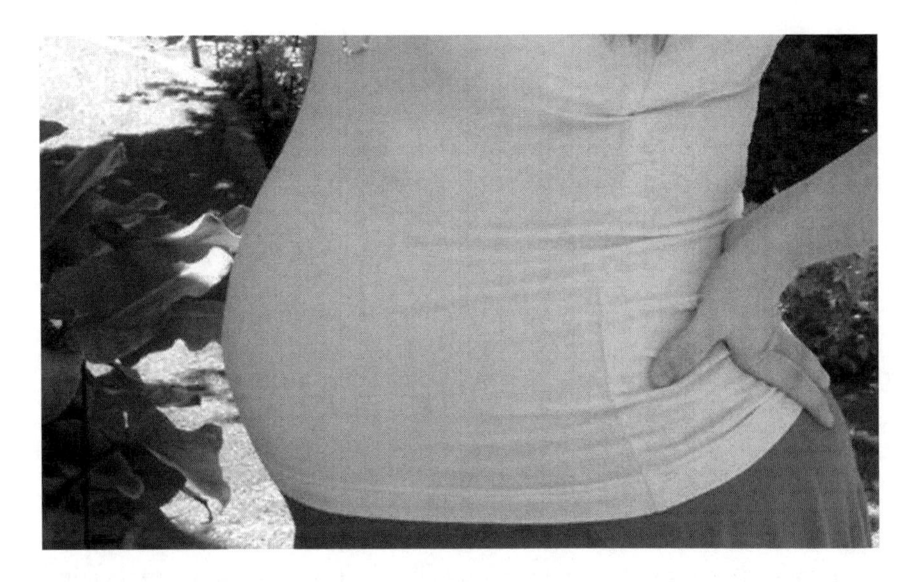

of view, this decline is bizarre. Other long-lived animals stay fertile until close to the end of their lives, with elephants breeding until their 60s and the great whales doing so in their 90s. In comparison, a human woman is exceptional in losing her child-bearing potential years or decades before losing her life. Even in hunter-gatherer societies that lack our access to modern medicine and technology, women who pass through menopause can expect to live well into their sixties. Now, a pair of scientists have proposed a new model to explain the origins of menopause. Michael Cant from the University of Exeter and Rufus Johnstone from the University of Cambridge suggest that the loss of fertility helps to lessen reproductive conflicts between successive generations of women.[61]

A few theories have already been put forward to resolve this conundrum. A recent explanation suggested that the menopause reduces the health risks that repeated childbirth brings to both mother and child. Simply put, there are only so many children that a woman can have before the process of childbirth puts her body and that of her child at risk. This idea complements the most popular theory, known as the "grandmother hypothesis", which suggests that older, infertile women can still boost their reproductive legacy by feeding, teaching and caring for their existing children and grandchildren. The basic idea makes sense and while some studies have backed it up, it is clearly not the whole story. Some analyses of hunter-gatherer populations have found that the indirect advantages of helping your family do not outweigh the potential benefits of having more children yourself. Alone, the grandmother hypothesis can explain why women continue to live past the menopause, but not why they go through it in the first place. Cant and Johnstone believe that the current picture is incomplete because previous studies have ignored the fact that new children affect not just their mothers, but other members of the community too. The children of all the fertile women within any group draw upon the same pool of food, resources and attention from other adults, which effectively leads to a form of "reproductive competition" between mothers.

Cant and Johnstone suggest that menopause serves to minimise this conflict and cite the timing of menopause as evidence for their theory. In humans, there is remarkably little overlap between the reproductive periods of different generations. In hunter-gatherer societies, mothers tend to stop being fertile at more or less the same time that their daughters become sexually mature. This degree of separation is truly exceptional among other primates, which mostly become fertile while their mothers are still more than capable of conceiving. For example, the fertile periods of successive generations of Japanese macaques overlap by about 12 years, which is about 70% of their total reproductive lifespan. Based on the trends shown by other primates, human women would be expected to keep on bearing children till the ripe old age of 70, rather than the much earlier cut-off in their 50s. From birth, a woman is outfitted with a lifetime supply of follicles, shells of cells that contain immature egg cells, and these are used up as she goes through more and more menstrual cycles. The stocks are gradually worn away but the process accelerates dramatically at about the age of 38, in a way that does not happen in chimps, monkeys or rodents. If this acceleration never happened, the earlier and slower rate of follicle loss would lead to menopause at around the age of 70, the same age predicted by trends in other primates.

If older and younger women do indeed experience reproductive conflict, why is it the older generation who cedes ground by becoming infertile, and not the younger one? After all, in most mammals that cooperate to raise young, it is the other way round – the older generation continues to breed and suppresses the fecundity of the younger generation. Cant and Johnstone believe that it all boils down to how the social groups of our ancestors mingled with one another. In all social mammals, groups exchange members to some extent and in most cases, it is the males that strike out. But for ancestral humans, there is plenty of evidence to suggest that the females were more often the ones who left home and found new groups. Genetic evidence, along with the behaviour of hunter-gatherers and our close relatives, the chimps and bonobos, all support this idea. This simple fact changes the degree that different females are related to other members of their group and that shifts the balance of conflict in favour of the young newcomers. A young female entering a group is better off raising children of her own, for she is completely unrelated to the group's existing members and gains no indirect benefits from helping to raise their children. On the other hand, an old female can benefit from either having more children herself or helping to raise any grandchildren that her sons father with the young newcomers.

Using a simple model to simulate these interactions, Cant and Johnstone found that these asymmetric benefits skew the results of the competition towards the younger females. The competition resolves itself in a stable way if the older females stop reproducing when the younger ones begin. The duo freely admit that their hypothesis will need to be tested further and suggest ways of doing so. For a start, they note that if they are correct, scientists should be able to show that young mothers experience drawbacks if they have children alongside older grandmothers who are still doing the same, as sometimes happens in polygamous societies. Finally, Cant and Johnstone note that their new hypothesis is not meant to be an alternative to existing ones, but a complement to them. They hope that it will help us to more fully understand the origins of menopause if we view it as a reflection of the "ghost of reproductive competition past".

The positive side of inbreeding

Marriage between closely related cousins is a heavy taboo in many cultures and its critics often cite the high risk of genetic diseases that inbreeding brings. That risk is certainly apparent for very close relatives, but a new study from Iceland shows that very distant relatives do not have it easy either. In the long run, they have just as few children and grandchildren as closely related ones. Sex chromosomes aside, every person has two copies of each gene, one inherited from their father and one by their mother. Not every gene will be in correct working order, but there is a good chance that a faulty copy will be offset by a functional one from the other parent. However, if two parents are closely related, there is a higher-than-average chance that they will already share some of the same faulty genes and a similarly increased chance that their child will receive two defective copies. That can be very bad news indeed and in cases where important genes are affected, the results can include miscarriage, birth defects or early death. Sex, then, is a shuffling of the genetic deck and theoretically the more closely related that partners are, the greater the chance that their child will be dealt a dud hand. And yet, some studies have found that some closely related couples actually do better than distant relatives in terms of the number of children they manage to raise. This trend is certainly unexpected and the big question is whether it is the result of biology or money.

In societies where close relatives marry, these unions tend to happen at a relatively early age and they provide avenues for families to retain wealth and land within bloodlines. These related couples enjoy the health benefits enjoyed by the rich as well as more time in which to raise a larger family. Together, these two effects could more than make up for any disadvantages wrought by their genes. Earlier studies have done little to clear the confusion. They have mostly been conducted in parts of the world like India, Pakistan and the Middle East where marriage between close relatives is relatively frequent, but which are also home to enormous gulfs between the richest and poorest members of society. With demographics like these, sorting out the relative contribution of socioeconomics and biology is difficult. To do that, what you need is a country with a small population where couples are reasonably closely related and with a very shallow gradient between rich and poor. Ideally, you'd also want this country to have excellent family records dating back several years. In short, you'd want to base your study in a country almost exactly like Iceland.

Iceland is home to a tiny population of just over 300,000 people who enjoy a level of social equality that is almost unparalleled elsewhere in the world. Wealth, family size and cultural practices are fairly uniform. The country is also home to uniquely impressive genealogical records that allow today's Icelanders to track their family trees with exacting precision for centuries. These records are supplemented by thorough medical records and thousands of willingly donated genetic samples. Agnar Helgason from deCODE Genetics, a pharmaceutical company located in Reykjavik, made good use of these records to study over 160,000 Icelandic couples since 1800.[62] At this time, Iceland was still a poor agricultural nation and close-knit rural communities meant that on average, couples were related at the level of third or fourth cousins. Since then, the country has prospered into a wealthy industrial one and the growing population has shifted to a mainly urban way of life. In doing so, people became more likely to find partners who were more distantly related and by 1965, couples were only related at the level of fifth cousins on average.

As expected, Helgarson's study unveiled the dangers of close inbreeding. While the most closely related couples had the highest number of children, many of them failed to live long enough to have children of their own and in the long run these couples had the fewest grandchildren. But surprisingly, distantly related couples were at a disadvantage too. In fact, Helgarson found that couples related at the level of third cousins eventually fostered the largest families. For example, among women born between 1800 and 1824, those partnered with men who were third cousins had an average of four children and nine grandchildren, while those partnered with a distant eighth cousin had just three children and seven grandchildren. For starting large families, very distant relatives were just as poor prospects as very close ones. Over the 200 years included in the study, Iceland has seen a steep decline in both fertility and relatedness between couples. And despite all that, for every 25-year period that Helgarson looked at, the same pattern held – couples who were moderately closely related ended up with the largest number of descendants.

These remarkably consistent results have convinced Helgarson that the counterintuitive effect must have some biological foundation. Its exact nature will have to wait for another study and for now, we are left only with speculation. It could be that a child's immune system may be more compatible with its mother's if its father is reasonably closely related to her. Alternatively, a union between distant relatives could serve to splinter groups of beneficial genes that have evolved in close association with each other. The study's implications for societal taboos against marriages between close cousins is open for debate. Certainly, it does not mean that singletons should be sifting through their address books on the hunt for attractive third cousins. However, the relatively poor reproductive success of distant relatives has the potential to explain the massive decline in fertility in many countries the world over. In the time that Iceland has gone from rural agriculture to urban industry, its population growth has slowed and its fertility rates have declined, a trend shared by a slew of other nations. Helgarson suggests that this could, at least in part, be due to people finding ever more distantly related partners.

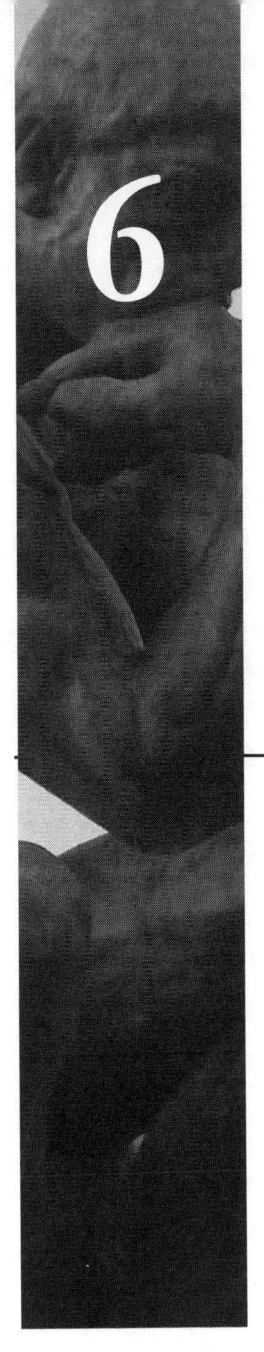

Rethinking the familiar

Why it pays to make refs see red, why you are not really undecided and how to buy happiness

Referees see red

For a sportsman, sometimes it pays to have the referee seeing red. In some sports, a simple red garment can give an athlete a competitive advantage because the striking hues draw the focus of the referee. With a delightfully simple but beautifully crafted experiment, Norbert Hagemann at the University of Munster found that refs have a tendency to award more points to red-garbed competitors.[63]

Three years ago, Russell Hill and Robert Barton found that in boxing, tae kwon do and wrestling, contestants who wore red are more likely to win their bouts. They reasoned that red – the colour of anger and aggression – gives athletes who don it a psychological boost that brings victory that much closer. An interesting idea, but a later study by Candy Rowe showed that there is nothing special about red. She found that in judo, where players wear either blue or white suits, it is blue that grants the advantage. Her theory was that white suits are brighter than the blue ones and more strongly contrasted against the background. That gives the blue competitors a edge when it comes to following and anticipating their opponents' moves. But Hagemann rejects both ideas. His explanation has nothing to do with the players at all, and everything to do with people who do not take part but wield massive influence over the outcomes of competitive sports – the referees. They are frequently forced to make very difficult decisions under less-than-ideal conditions. Under these tough circumstances, small biases in their perceptions – like a tendency to look at the more striking colour – may come into play.

To test this idea, Hagemann and colleagues recruited 42 seasoned referees who, between them, had over 320 years of experience at judging tae kwon do matches. The judges watched one of two sets of videos clips, showing matches between contestants who wore red or blue gear. Both sets of clips were completely identical save for one key difference – in one, the experimenters had digitally swapped the colours of the fighters' clothes. The referees watched the matches and, following the rules of the World Taekwondo Federation, awarded their points. And when Hagemann tallied up the scores, he found that the red fighters scored on average 13% more points than the blue ones. When a blue competitor was digitally transformed into a red one, his points tended to rise but a red competitor who was turned into a blue one was awarded lower scores. Neither set of clips was given a higher total number of points. It is a very neat result – with a bit of digital sleight-of-hand, Hagemann found that referees will assign more points to red-clothed contestants than they would to blue-clothed ones, *even if their performance is exactly the same.*

It is unclear whether this referee bias explains Rowe's judo results, especially since she suggested that white is the more vivid colour. But Hagemann feels that it can certainly explain the Hill and Barton study where red competitors had a slight edge over white. In their analysis, the edge bestowed by crimson hues only mattered when contestants were closely matched because only then are the referees' decisions important enough to tip the scales towards the more vivid contender. If one contestant is clearly superior, referee bias is not going to make much of a difference even if one of the pair is wearing yellow with purple polka dots. It is also worth noting that this referee effect only really applies to combat sports like tae kwon do and boxing where it is the ref who actively awards points. In these cases, it may be a good idea to change the rules so that red attire is not allowed, or that referee decisions are supplemented with electronic aids. In team sports like football, where the referee's role is more to do with regulating fair play and punishing foul moves, it is unlikely to have much effect, so Chelsea need not worry when playing Arsenal. Indeed, in sports like football, it may even be counter-productive to draw the referee's attention if your primary strategy is to dive towards the pitch in mock agony whenever another player glances in your general direction.

Undecided voters not really undecided

Elections are weighing heavily on our minds. By the time this book is published, the race between Barack Obama and John McCain in the US will probably have come to a head, while Britain will be looking forward to a general election within the next few years. Some people, of course, will vote based on long-held loyalties to a specific political party, but many are more malleable in their picks. What affects the choices of these undecided voters? People are given to viewing ourselves as rational beings and as such, we'd like to think that our choices are fuelled by objective and careful deliberation. So we pay attention to media coverage, we read up on policies and we listen to debates and only then, having gathered as much information as we can about the various options, do we make a choice. That is how it plays out in our heads, but according to a new study, the reality may be quite different. Silvia Galdi at the University of Padova, Italy, has found evidence that the final verdicts of undecided decision-makers are only weakly related to their conscious preferences and more strongly influenced by unconscious views and biases they are not aware of.[64] In many cases, when people claim that they are undecided, they have secretly made up our minds, unbeknownst even to themselves. For example, a British voter sitting on the fence might unconsciously be inclined to vote for David Cameron because they view Gordon Brown as dour, or oppositely because of a prejudice against the Tory party. Likewise, and more unfortunately, an American voter might side with John McCain because of unconscious racial prejudices against black people.

By their very nature, there unconscious associations are not easy to find, but psychologists have a tool for doing so – implicit association tests. Volunteers are shown a series of words or images and must classify them into one of two categories by pressing assigned keys. For example, they might have to distinguish good words (happy, joy) from bad words (anger, hate) and white faces from black faces. At first, the categories are presented separately and then in various combinations. So in one trial, you might be asked to press one key for good words and black faces and the other key for bad words and white faces. The idea is that people perform the task more quickly and more accurately if the combinations of categories matches their unconscious associations. So if people have a hidden prejudice against black people, they would be quicker at trials where black faces and bad words were represented by the same key, than those where black faces were twinned with good words. If you want to see these tests in action, Harvard have a large range of them online and I'd highly recommend having a go yourself (https://implicit.harvard.edu/).

Galdi unveiled the true influence of these hidden biases by interviewing 129 residents of Vicenza, Italy about the expansion of a new US army base nearby. At the time, the expansion was a burning issue in the media and strongly polarised the opinions of local Italians. Galdi asked all the interviewees outright about whether they were in favour of enlarging the base, against it or undecided. She also asked them a set of 10 questions that probed deeper into the conscious reasons behind their decisions, be they environmental, economic, social or political. So much for their conscious beliefs. To bring out their unconscious ones, Galdi gave them a variation of the implicit association test where they had to classify pictures of the base, as well as positive and negative words, as quickly as possible. All the interviewees returned to repeat the tests one week later. Among those who were previously undecided about the base, Galdi found that their conscious beliefs had little bearing on their later choices. Instead, it was their unconscious biases that had the greater influence; they predicted which way the interviewees' decisions would swing a week later, as well as any changes in the direction of their conscious beliefs.

The tests show that the unconscious beliefs of these swing-voters were influential enough to sway their future decisions. Even though they said (and most probably believed) that they were undecided, they had to some extent already made up their minds. People who had already made up their minds behaved differently. In their brains, unconscious associations held little sway and it was their conscious reasons that predicted their future choices. In fact, these reasons even predicted any changes in their unconscious associations – their beliefs were strong enough that over time, they eventually strengthened into a sort of mental reflex. So the minds of the decided resolutely stay their course, but those of the undecided are surprisingly affected in ways they are unaware of.

But what then of the painstaking deliberation process that many people go through to make decisions? Is there any point to lists of pros and cons, a reading of reviews, or a careful gathering of balanced viewpoints? Based on other studies, Galdi suggests that in many cases, these acts merely serve to confirm and support decisions that have already been unconsciously made. The idea is that your inner biases affect which bits of information you pay attention to, and they affect the way you interpret any data you do take in. Your secret preferences for candidate A over candidate B (and they are secret even to yourself) predisposes how you process new information in a way that favours A over B. In time, your unconscious favouritism becomes a conscious preference but your Eureka moment is the result of a long-term manipulation by silent puppet-masters.

These are lessons that political pundits and pollsters might care to heed. As elections near, vast acres of forest are felled in order to print the results of opinion polls and untold amounts of glucose are burned by analysts poring over the results in a vain attempt to understand why people voted the way they did. But surely, this picture cannot be a complete one. It relies solely on the conscious reasons that people offer during interviews, reasons that we now appreciate are often elaborate fictions. And people who claim they are undecided may well be telling the truth as far as they are concerned, but actually be far from it. You might argue that the interviewees were just deliberately trying to hide their views and biases, especially if they were unpopular or taboo. Certainly, it is a valid alternative way of construing the results. But Galdi's interpretation does not stand alone – it sits alongside a slew of other studies, which show that we are often in the dark about our own decision making processes and how they are influenced.

Take the work of Petter Johansson's group on "choice blindness". In 2005, they asked a group of volunteers to choose which of two women they found most attractive based on photos. They were then shown their selected image and asked for the reasons behind their choice. But in some trials, the deft experimenters performed a sleight-of-hand that swapped the photos and presented the volunteers with the photo they had rejected. Amazingly, 75% of people failed to notice the switch. Even more amazingly, the duped volunteers had no problems in explaining the choices they did not actually make. "She is radiant," said one. "I like earrings," he continued. The most telling result was the fact that there weren't any differences between the volunteers' reasons for their real choices and the reasons for the choices they did not actually make. Both types of rationale contained the same level of detail and were expressed with the same confidence and emotion. Like the undecided voters in Galdi's study, the duped men in Johansson's were also unaware of their own unawareness and they happily justified their "preferences" to themselves. Studies like these show that conscious decision-making, as we know it, is just the tip of a psychological iceberg, with the bulk of the process operating out of view. We are often strangers to ourselves.

But why should people be so "unaware of our unawareness"? For a start, this hidden world of unconscious processing makes for a more efficient thinking machine by relieving us of the burden of micro-managing every trivial decision. It could also be that our mental lives already seem so rich and saturated with information that we find it hard to conceive of levels of processing that we cannot see.

How to buy happiness

> "This planet has – or rather had – a problem, which was this: most of the people living on it were unhappy for pretty much of the time. Many solutions were suggested for this problem, but most of these were largely concerned with the movements of small green pieces of paper, which is odd because on the whole it was not the small green pieces of paper that were unhappy." – Douglas Adams

In this pithy paragraph, the sorely missed Douglas Adams sums up a puzzling paradox of modern life – we often link happiness to money and the spending of it, even though both proverbs and psychological surveys suggest that the two are unrelated. Across and within countries, income has an incredibly weak effect on happiness once people have enough to secure basic needs and standards of living. Once people are lifted out of abject poverty and thrown into the middle class, any extra earnings do little to improve their *joie de vivre*. Time trends tell a similar story; even developed countries that have enjoyed economic booms have seen plateauing levels of satisfaction. But a new study reveals that money can indeed buy happiness... if it is spent on others. Elizabeth Dunn from the University of British Columbia wanted to see if there were ways of channelling the inevitable pursuit of money towards actually making people happier. Together with Lara Aknin and Michael Norton, she asked a representative group of 632 Americans to disclose their average monthly expenditure and to rate how happy they were.[65] She found that personal

spending, including bills, living expenses and treats for oneself, made up 90% of the average outgoings but had no bearing on satisfaction. On the other hand, people who spent more money on others by way of gifts or charitable donations, were much happier for it. That either suggests that selfless spending increases happiness, or just that happier people are more likely to plump up more money for friends or charities.

Dunn sought out firmer conclusions by watching what happened to people who received an unexpected windfall. She surveyed 16 employees at a Boston firm who were given a bonus that ranged from $3,000 to $8,000. About two months later, Dunn grilled them about how they had spent the money and again, regardless of the size of the bonus, those who devoted more of their windfalls to selfless ends ended up happier, while those who splashed out on themselves did not. To paraphrase a saying, it is not how much you have, it is what you do with it that counts. Finally, Dunn tested this theory through an experiment. She gave 46 people either $5 or $20, and an afternoon to spend it. Half of the lucky volunteers were told to splurge on themselves, while the other half had to buy a gift for someone else, or to give the money to charity. By the evening, the charitable individuals felt happier than they did in the morning while the self-spenders did not, regardless of which bill they were given, and despite the fact that they were acting on instructions.

Dunn's results have far-reaching implications. For a start, they suggest that many people are seeking happiness by spending money in the way that is least likely to actually make them happy in the long run – chasing after expensive consumer goods that will give them a mere temporary fix of pleasure. The modern obsession with personal spending is rather like running on a hedonic treadmill. And in a *deeply* ironic twist, the types of behaviour that allow money to buy happiness are subverted by the presence of, you guessed it, more money! Higher incomes bring greater self-sufficiency and as people start to need less help themselves, they tend to provide less for others. In some psychological experiments, just the mere *thought* of money can make people less likely to donate to charity, help acquaintances or spend time with friends, exactly the types of behaviour that are linked to happiness.

An emerging viewpoint from the science of happiness is that a person's circumstances in life – their income, jobs and so on – tend to have limited long-term effects on their happiness. People mentally adapt to stable situations unless they learn to actively engage with their circumstances – simply put, savour the moment or your goalposts will shift. This latest study is consistent with this idea, for it showed that the way in which money is spent has a greater bearing on contentment than how much is made.

There is a silver lining then. While Dunn's work implies that of selfless spending is the key to happiness, it also suggests that you do not need to pauperise yourself to do it. The experimental study suggested that paying as little $5 towards a selfless cause can result in a significant spike in happiness. Given that the volunteers in the first study only spent 10% of their earnings on other people, there is plenty of leeway for purchasing a bit of pleasure. And if all of that seems obvious in hindsight, consider this: when Dunn asked a fresh group of 109 people about the things that would make them happiest, she found that they were, on average, doubly wrong. A majority of 63% predicted that personal spending would make them happier than selfless spending while 86% said that they would be happier with the $20 bill than the $5 one. Those are certainly the intuitive answers, but they are not the empirical ones. Which would you believe?

Why studying will not help you remember

It's a familiar scene – the wee hours of the morning are ticking away and your head is bent over a stack of notes, desperately trying to cram as much knowledge in before the test in the morning. Because of the way our education system works, this process of hard studying has become almost synonymous with the act of learning, and the inevitable tests and exams that bookend this ordeal merely assess how much information has stuck. But a new study reveals that the tests themselves do more good for our ability to learn that the many hours before them spent relentlessly poring over notes and textbook. The act of repeatedly retrieving and using learned information drives memories into long-term storage, while repetitive revision produced almost no benefits.

To separate the effects of studying and testing on memory, Jeffrey Karpicke from Purdue University and Henry Roediger III from Washington University in St Louis set a group of local university students to a simple learning exercise.[66] At first, all of them were asked to study 40 pairs of English words and their Swahili equivalents, such as boat and *mashua*, and were then tested on each pair to see how many they could remember. The students were then split into four groups who were put through three more rounds of studying and testing in slightly different ways. For Group One, the three further rounds were exactly like the first. For Group Two, any pairs of words that they got right were dropped from the study lists but were still tested. Group Three faced the opposite condition – their correct answers were dropped from the subsequent tests but they were still asked to study them. And finally, Group Four's correct answers were omitted both from later study sessions and later tests. This fourth situation most closely mirrors what conventional wisdom tells us to do. Once something is learned, that is the end of the story and our attentions should focus on trickier facts that have not been so amenable to memory.

After four rounds, the students had done an almost perfect job. Almost all of them remembered almost all of the 40 word pairs and all four groups picked up the words at the same rate. The successful students were sent away, but Karpicke and Roediger reconvened them one week later for a final test of their rudimentary Swahili vocabularies. All the students from the four groups had predicted that in a week's time, they would remember about half of the words they had learned. But the reality was very different – two of the groups achieved much higher scores than the others. Repeated testing was the critical factor. The first two groups were repeatedly tested on all the word pairs regardless of whether they successfully remembered them or not. A week later, they still remembered 80% of the words they had learned and Group 2, who did not have to study words they correctly remembered did just as well as Group 1, who had to pore over the full lists. Students in the last two groups were not tested on words once they had remembered them correctly once, and they suffered for it. After a week, they remembered only about 30% of the 40 pairs. Even Group 3, who repeatedly studied every pair in every study period fared much worse in the long run than Group 2, who studied selectively but were tested comprehensively. The study's results are as amazing as they are counter-intuitive. They showed that where long-term memory is concerned, the act of repeated studying brings essentially no benefits, especially once a piece of information can already be dredged up from memory. Repeated testing, on the other hand, is of paramount importance, even for facts that can already be successfully recalled.

If this seems like an obvious conclusion in retrospect, consider the fact that the students themselves were unaware of it. All of them predicted that they would score about 50% in the final test where in fact, the average scores were either substantially higher or lower. Indeed, surveys have shown that very few students use self-testing as a strategy for revision. When they do, it is more to work out what they have or have not learned rather than as an active part of the process. And once they have successfully memorised something, they tend to drop it from further practice. Karpicke's and Roediger's study makes a case for tests and exams to move away from their crude use as assessment tools. Currently, they are stressful and high-stakes affairs, deployed at the end of academic terms to mark an endpoint of the learning process. Perhaps, if they were used in a more continuous and informal capacity, they could actually contribute to the process too.

Punishing slackers and do-gooders

Humans have an extraordinary capacity for selflessness. We often help complete strangers who are unrelated to us, who we may never meet again and who are unlikely to be able to return the favour. More and more, we are being asked to behave in selfless ways to further the common good, not least in the race to tackle climate change. Given these challenges, it is more important than ever to understand the roots of cooperative behaviour. From an evolutionary point of view, it can be a bit puzzling because any utopic society finds itself vulnerable to slackers, who can succeed at the expense of their peers while contributing little themselves. When cheaters can all-too-easily prosper, it should be difficult for altruistic behaviour to persist. Nonetheless, persist it does, and psychological experiments suggest that punishment is part of the glue that binds a cooperating society together. In general, we as a species value fair play and we loathe freeloaders, to the extent that we are all too willing to sacrifice our own gains in order to punish cheats. But punishment is a two-way street and not all freeloaders take castigation lightly. They can easily get their own back on altruists out of revenge or a simple desire to take down some do-gooders. This 'antisocial punishment' is often ignored by social science research but it actually has the ability to derail the high levels of cooperation that other fairer forms of punishment can help to entrench.

To study the links between punishment and cooperation, Benedikt Herrmann from the University of Nottingham watched the behaviour of university students from 16 cities around the world as they played a psychological game.[67] He picked cities as diverse as Boston, Copenhagen and Riyadh to compare behaviour across a wide range of cultures, but worked with upper- or middle-class university students to study people from comparable social groups. The students played a 'public goods game' in groups of four, anonymously and over computers. Each was given 20 tokens and told to place a number of their choice into a public account. This joint money gained interest by a factor of 1.6 and was distributed equally between the players, after which each individual pot was converted into real money. The game went on for ten rounds and after each one, the players were informed about the moves that their peers made. From the groups' point of view, the best decision was for everyone to put in their full pot, and walk away with 32 tokens apiece. However, each individual player would do best by putting nothing in and nonetheless reaping a share of their peers' contributions – they would then receive 44 tokens for no personal risk. But if none of the players contributed anything, no one would profit and everyone would stay on 20 tokens apiece.

In reality, the groups showed very different degrees of cooperation, with players initially chipping in between 8-14 tokens from their total of 20. However, as they became aware of spongers in their midst, they became less motivated to chip in themselves and in all groups, the contributions dwindled over time until players were only putting in a measly average of five tokens each. Herrmann then repeated the experiment with a twist. After each round, players had the chance to anonymously punish their peers with a fine of up to 30 tokens, which were then removed from their pot. This action came at a cost, and each punisher had to pay one token for every three they removed from someone else. With this opportunity for enforcement, the situation changed dramatically. Students in all 16 groups were more than happy to sacrifice their own tokens to punish freeloaders, and all the groups did so in very similar ways, doling out the heaviest fines to those who contributed the fewest tokens to the pot. Faced with the imminent threat of punishment, freeloaders lost the impetus to cheat and none of the groups showed a breakdown in cooperation.

But Herrmann also saw substantial differences in the way that the 16 groups reacted to punishments. In some countries like the USA, the UK and Australia, admonished cheaters mostly accepted their punishment and started to donate more over time. As a result, cooperative behaviour became more common and by the end of the tenth round, many of the groups were chipping in about 70-90% of their pot. In contrast, some freeloaders took less well to having their wrists slapped and sought revenge on their punishers, regardless of how generous they had actually been. Indeed, players who were punished most heavily were those who were more likely to exact antisocial punishments on their peers. This practice of 'antisocial punishment' was largely absent in many groups but it was a frequent tactic among players from countries like Saudi Arabia, Greece and Oman, who were just as likely to castigate altruists as scroungers. Even in these groups, levels of cooperation did not decline but they never climbed to the levels of groups that only punished freeloaders. By the end of the tenth round, players from Athens and Riyadh were only contributing about 33% of their earnings.

These international differences in cooperation were reflected in the final earnings of the different groups. The most generous players (from Boston) walked away with over 2.5 times more money than the least generous ones (from Muscat). Herrmann found that antisocial punishment was more common in countries where the ethic of cooperation is less ingrained. He measured the attitudes of the 16 countries included in the game, using data from the World Values Survey, which asked people about their views on whether activities like tax evasion, benefit fraud and fare-dodging are ever justified. The 16 scores spanned almost the entire worldwide range of attitudes, and Herrmann found that countries that show greater tolerance for these freeloading actions were more likely to exact antisocial punishment. Conversely, he reasoned that societies with strongly ingrained views on the value of cooperation are more likely to abhor free-riding, applaud altruism and avoid antisocial punishment. Herrmann also found that antisocial punishment was less common in countries that trusted their courts, police and other law enforcement officials to be effective, fair and free from corruption. On the other hand, if people believed that the rule of law was weak (as measured by the World Bank's "Rule of Law" index) they were more likely to seek revenge against those they felt had wronged them. In such situations, antisocial punishment was not only more common but meted out more harshly.

As Herrmann's work shows, the degree of antisocial punishment in a society can have a strong impact on how cooperative it is and in turn, how economically successful it is. The results of the game suggest that a country's economic prowess depends not just on a naked desire for material gain, but also on its moral attitude towards cooperation. It is tempting to attribute this ethic of cooperation to democracy and indeed, six of the countries that showed the lowest levels of antisocial punishment also score highly in rankings of civil liberties, political rights and freedom of the press. Authoritarian societies tend to show higher levels of antisocial punishment but this delineation is perhaps too simplistic. China, for example, strongly bucks the trend in that it scores poorly in terms of the democratic markers listed above, but still shows relatively low levels of antisocial punishment and accordingly, is one of the world's fastest growing economies.

There is, however, another twist to the story. In a separate study, Anna Dreber, Martin Nowak and colleagues from Harvard University confirm that groups of people are indeed more likely to cooperate if they can dole out punishment, but they also reap smaller rewards.[68] In their experiments, the groups that left with the highest payoffs were those that shunned punishment completely. It is a conclusion best summed up by the stark and simple title of their paper: "Winners do not punish." Dreber revealed the dark side of punishment by modifying one of the classic experiments of game theory – the Prisoner's Dilemma. The game is inspired by the plight of separately interrogated prisoners, who are given the option to rat the other one out in exchange for immunity. If both stand firm, they walk away, but if one holds their ground while the other gives in, he gets done for the crime. The game version of this dilemma pits two players against each other, and gives them a choice to defect or cooperate. For each 'prisoner', the best choice no matter what his partner does is to defect, but if they both defect, their outcomes are far poorer than if they had both cooperated – hence the dilemma. The artificial scenario represents many real-world choices where cooperation is good for a group but cheating is best for the individual. When defection presents such stark benefits, evolutionary theory predicts that it should be commonplace unless some force can maintain cooperation. Recently, costly punishment – where an individual can suffer slightly to punish a cheater – has been mooted as such as force, because people avoid cheating for fear of reprisals.

To test this idea, Dreber extended the Prisoner's Dilemma so that on every turn, players could punish as well as cooperate or defect. She recruited 104 local college students to play anonymously against each other. They were given a set of virtual tokens, informed about their options in neutral language (so as not to influence their behaviour) and asked to make their moves together. Once played, the results were tallied and the choices revealed. The games went on for different times but players knew that every round had a one in four chance of being the last one. At the end of each game, each player was paid a dime for every remaining point.

In one game (T1), cooperation meant paying one token for the other player to receive 2, defection meant taking one token from the other player and adding it to your pot, and punishment meant paying one unit to fine the other person 4. The second game (T2) was exactly the same, except that cooperation was more valuable, with the beneficiaries receiving three tokens instead of 2. Alongside T1 and T2, Dreber also ran two control experiments (C1 and C2) where players could only cooperate or defect and punishment was not an option. As the experiments unfolded, a number of different

	Event	Moves	Payoffs
A	Nice people finish first	C C C C	8
		C C C C	8
B	Punish and perish	C P P P P	-10
		D D D D	-9
c	Defection restores cooperation	C D D C D C	10
		D D C C C C	4
d	Turning the other cheek	C C C C C	2
		D D C C C	14
e	Mutually assured destruction	C P P P D D	-20
		D D P P P P	-14
g	A 'pre-emptive strike'	C P D	2
		C C D	-4

Table 2. Examples of decisions made during real Prisoner's Dilemma games. Players can either cooperate (C), defect (D) or punish (P).

strategies became apparent. Some games were all-out cooperation (a). In others, one player defected but cooperation was restored when the other player either turned the other cheek (d) or defected in retaliation (c). In the face of forgiveness or the threat of mutual loss, the original defector decided to play fair again. When punishment was played, it usually did not restore cooperation. In some cases, the rebuked player simply carried on defecting only to be punished even further (b). When a punished player retaliated in kind (antisocial punishment), the game ended in mutually assured destruction (e). Finally, the ability to punish allowed irrational individuals to inflict harm on the undeserving with unprovoked pre-emptive strikes that had disastrous results for cooperation (f).

Even though actually taking punitive measures proved to be an anathema to teamwork, the option to punish *did* increase the overall levels of cooperation. In the two games that allowed punishment, T1 and T2, players chose to cooperate in 52% and 60% of their moves, but they only did so in 21% and 43% of the moves in the punishment-free control games, C1 and C2. That seems like a good case for punishment, but not so. Dreber found no difference between the average takings in the two setups that included punishment (T1 and T2) and the two that did not (C1 and C2). As far as the groups were concerned, the ability to punish bore no benefits. At an individual level, things were even worse, for the players who came away with the least money were also those who meted out punishment most frequently. In the T1 game, for example, the five players who ended up richest were those who *never* punished their opponents. You might suspect that these winners were just lucky and only faced opponents who always cooperated and never deserved punishment. But that was not the case – their opponents occasionally defected too and the one strategic choice that set the winners apart from the losers was how they dealt with a defection.

Simply put, losers chose to punish while winners opted for a 'tit-for-tat' strategy and defected themselves. Winners, it seems, really do not punish.

Hermann's study actually found similar patterns. In it, players showed more cooperative behaviour when they were allowed to punish their colleagues than when they weren't. But it also found that 13 of these groups actually ended up with lower average earnings in the games that involved punishment than those that did not. Only three groups netted higher payoffs and Dreber suspects that the differences were not statistically significant. She said, "I believe that our results agree with those of Hermann et al.: punishment leads to more cooperation, but not higher payoffs." The results are a blow for the notion that costly punishment was critical for the evolution of human cooperation, for people who resort to punishment suffer for it. Instead, the authors suggest that costly punishment may have evolved for other reasons, like establishing pecking orders or allowing stronger members of a group to dominate weaker ones through coercion. As Dreber herself says:

> "People engage in conflicts and know that conflicts can carry costs. Costly punishment serves to escalate conflicts, not to moderate them. [It] might force people to submit, but not to cooperate... Winners do not use costly punishment, whereas losers punish and perish."

The power of powerlessness

Feeling powerless is no fun. A lack of control can make the difference between contented and unhappy employees. But a lack of power does not just make people feel disgruntled; it has a more fundamental effect on their mental skills. In a series of experiments, Pamela Smith from Radboud University Nijmegen has shown that the powerless actually take a measurable hit to important mental abilities.[69] Even if people are subconsciously primed with the *concept* of being powerless, they perform more poorly at tasks designed to assess their ability to plan, focus on goals and ignore distractions. According to previous research, a lack of power forces people to constantly re-evaluate their own goals and monitor more senior individuals. Without authority, a person's actions rely on instructions and may constantly change at the whim of their superiors, whose own motives and goals must be guessed at.

Monkeys show similar behaviour. Studies have found that subordinate rhesus males follow the gaze of those with higher status, while dominant

individuals only look in the same direction as others with even greater standing. Smith reasoned that this constant re-evaluation draws the brain's resources away from other needs, including a set of mental abilities known as "executive functions". The term is loosely defined but accurately named and refers to a set of master processes that control more basic abilities, like attention and motor skills. They allow us to plan for the future, adapt to new situations and carry out our goals. They allow us to carry out actions that further our goals while restraining us from those that hamper them.

Smith used a variety of psychological tests to investigate how power, or a lack of it, affects executive function. The first of these, a "two-back task", involved watching a sequence of letters flash by and saying if the current letter matched the one shown two cycles ago. The test requires people to continuously work out if new information is relevant to a set goal. About 100 volunteers took the test and Smith split them into superiors and subordinates. All were told that they would do a practice run, then work together with the superiors in control. After the test, the subordinates would be paid according to evaluations from the superiors, who would themselves receive a fixed payment. In reality, the "practice run" was the real deal and as predicted, the subordinates made more errors than their powerful superiors. It was an interesting first result but did not account for the possibility that the subordinates were just preoccupied with their impending evaluation. The second experiment was subtler. No specific roles of authority were dished out. Instead, Smith subliminally influenced the participants' feelings of power by asking them to unscramble a set of innocuous sentences. The scrambled words included those either related to power (*authority* or *dominate*) or the lack of it (*obey* or *subordinate*). After this psychological "priming", the volunteers did a Stroop test, designed to ascertain their ability to suppress irrelevant responses that will not help them to achieve a goal. The goal was to say if different words were written in red or blue ink. If the words were unrelated to colour, the power-priming had no effect on their performance. Likewise, there was no effect if the words were colours written in matching ink (e.g. "red" written in red). In both these conditions, volunteers found it easy to focus on the task at hand. However, if the word and ink were mismatched (e.g. "blue" written in red), the test became more difficult and the powerless group made more errors.

The third experiment built on the skills needed in the previous tests and looked at the more complex executive ability of planning. Smith used the classic Tower of Hanoi game, where players have to move a pyramid of discs from one pole to another, one at a time, and without placing a larger disk atop a smaller one. The game is not straightforward (you can find versions to try online). The disks often have to be initially moved in the opposite direction to their final locations and this "sub-goal" conflicts with the bigger one of getting everything to the final place. Players have to be able to reconcile these competing priorities to make a workable strategy. Again, they were primed beforehand with feelings of high or low power by writing about a time when they had control over someone or when someone had control over them. And again, the low-power group performed more poorly than the high-power one, taking more moves to shift the disks.

The poorer performances of the powerless groups were not for lack of trying. Smith administered a series of questionnaires after the tests in which all volunteers reported making the same amounts of effort and were equally motivated to successfully complete the tests. The results have strong implications for both societies and businesses. The majority of us would want to avoid a world where social hierarchies are purely based on characteristics like gender or race. We would prefer to live in perfect meritocracies, where even people from disadvantaged groups can ascend to positions of power if they are talented high achievers. But this mindset is perhaps too facile; if we assume that the powerful are swimming in skill or motivation, it follows that the powerless are lacking in both. That certainly was not the case in Smith's experiments. The volunteers were split into different groups randomly rather than based on known skills or talent. Despite this, they still showed more powerful executive functions if they experienced power, or even the mere concept of it. These results suggest that poor performance of those that lack power does not provide sufficient evidence that power has been allocated fairly. An alternative explanation is that assigning someone a certain position can alter their mental skills in a way that confirms their standing. The powerful retain power because of the improved mental processes that it brings about, while impairments in the same processes keep people without power on the bottom rung. These effects make hierarchies incredibly stable, and lead the powerless into what Smith calls "a destiny of dispossession".

Power does not have to come through rank – it can also be assigned on a case-by-case basis. Delegating jobs to people is not simply about telling them what to do; good managers will hand over not only tasks, but the authority to carry out those tasks. Employees tend to perform better if they have the ability to make their own decisions within certain confines, and take ownership of their own work. The alternative – peering over their shoulders and dictating their every move – is a sure-fire route to poor results. The results are especially interesting for professions where stakes are high and errors can cost lives, such as medicine, or where critical events are rare and goals must be kept firmly in mind, such as security. In such situations, improving the executive functions of employees by awarding them greater self-sufficiency and authority seems to be a no-brainer.

Subliminal flagging

For all the millions that are poured into electoral campaigns, a voter's choice can be influenced by the subtlest of signals. Israeli scientists have found that even subliminal exposure to national flags can shift a person's political views and even who they vote for. They managed to affect the attitudes of volunteers to the Israeli-Palestine conflict by showing them the Israeli flag for just 16 thousandths of a second, barely long enough for the image to consciously register. These results are stunning – even for people right in the middle of the one of the modern age's most deep-rooted conflicts, the subconscious sight of a flag drew their sympathies towards the political centre. But in some ways, it is all that not surprising. The last decades of experimental psychology have shown us that the our conscious view of the world is a construct created by our brain. We simply cannot consciously process the barrage of information constantly arriving through our senses and to save us from a mental breakdown, our brain does a lot of subconscious computing. The upshot of this is that our decisions can be strongly influenced by sights, sounds and other stimuli that we're completely unaware of.

Our political views are no different. In an ideal world, we would base them on a rational consideration of the relevant facts and our own beliefs, but in the real one, subliminal symbols pull on the puppet-strings too. National flags should be capable of this; to many people, they carry a weighty importance out of all proportion to their nature as rectangular sheets of cloth. Ran Hassin from Hebrew University and Melissa Ferguson from Cornell University have clearly demonstrated this by showing

subliminal images of flags to Israeli volunteers.[70] Their partners, Daniella Shidlovski and Tamar Gross, asked 53 people about how strongly they identified with Israeli nationalism and how being an Israeli affected their identity. According to their responses, they were separated into a High group or a Low one based on their penchant for nationalism.

They were then asked to answer on-screen questions, half of which were about the Israeli-Palestine conflict. The answers worked on a scale from one to nine, with nine reflecting the most strongly nationalistic attitudes. Before the questions came up on screen, the researchers briefly showed the volunteers an image of the Israeli flag or a "control flag" that had a jumbled collection of the same blue stripes and lines that make up the normal flag. The flags flashed up so quickly that none of the volunteers saw it, and none of them reported doing so when they were explicitly quizzed about it later. When they were shown the control flag, the High group, as expected, answered the on-screen questions with a nationalistic bent, averaging a score of 6. The Low group's average score was closer to 2.5. However, when both groups saw the subliminal Israeli flag, they both converged to a middle-ground score of 4. Sixteen milliseconds of exposure to the flag was enough to close the ideological gap between the two groups. In a second experiment, Hassin showed that the flashed flag had the same moderating effect on opinions about Jewish settlers in the West Bank and Gaza. At this point, it is worth noting that the flags did not wield an irresistible mind-altering power. There was still variation in the volunteer's choices, but the average trend changed in a statistically significant way.

In a final test, Hassin and Ferguson repeated their experiment in a real-world setting, before and after a local election. They worked with 101 new volunteers and asked them to reveal who they were planning on voting for before the event, and who they eventually voted for. The candidates were given a score from one to 6, with higher scores reflecting right-wing stances and lower ones reflecting left-wing ones. Amazingly, they found the same effect. Volunteers who saw the real subliminal real flags, but not the control ones, claimed that they were more likely to vote for the centre candidates, and actually did so. It is a shocking testament to the power of subliminal imagery – a quick flash in a laboratory can prime a person's behaviour some time later. It can even affect the most important political action of all – voting. Hassin's group note that they have uncovered a fascinating phenomenon but they are still in the dark about how it works. Do the symbols affect the weight we give to different views or do they affect our innate biases? Why does the Israeli flag drive people towards the political centre, and would symbols used by more extreme ideologies shift political stances away from it? The answers to the questions will have to wait. For now, the study serves to reiterate how important a simple symbol can be.

The Matrix illusion

In *The Matrix*, when Neo is first confronted by an Agent, his perception of time slows down, allowing him to see and dodge the oncoming bullets. Back in the real world, almost all of us have experienced moments of crisis when time seems to slow to a crawl, be it a crashing car, an incoming fist, or a falling valuable. But we now know that this effect is an illusion. When danger looms, we do not actually experience events in slow motion. Instead, our brains just remember time moving more slowly after the event has passed. Chess Stetson, Matthew Fiesta and David Eagleman demonstrated the illusion by putting a group of volunteers through 150 terrifying feet of free-fall.[71] They wanted to see if the fearful plummet allowed them to successfully complete a task that was only possible if time actually moved more slowly to their eyes.

The task was deceptively simple. They merely had to read two numbers that were displayed on a wrist-mounted machine called a 'perceptual chronometer'. Like a clunky digital watch, the device was programmed to show two numbers, but the catch was that the glowing digits were rapidly alternated with their negative images, where the number is dark and the area around it is lit. As the two images flicker more and more quickly, there comes a sudden point where they blur into a single uniform square of light. At this point, the rush of visual information overwhelms the brain of the volunteer, who is unable to resolve the two images apart. The trio of researchers tuned the device to each volunteer's threshold of resolution – the point where they only just failed to read the numbers. They reasoned that if a scary experience really made time slow down for the volunteers, even by a tiny amount, the flickering numbers should slow down enough to pop out of the blur. The effect should be like a slow motion camera, resolving the blur of a buzzing fly into individual wing beats.

To provide the necessary fear, Stetson performed the type of experiment that most scientists can only dream about. He took his volunteers up a SCAD tower (Suspended Catch Air Device) where they were strapped to a harness and dropped from a height of 150 ft onto a safety net. As they plummeted in free-fall, they had to try and read the numbers flashing from their wrists, while an eagle-eyed experimenter watched from the top to rule out those who kept their eyes completely shut. The volunteers failed. In fact, they read the numbers just as inaccurately as a control group who did the same task while staying on the ground. Unlike the slowed bullet-time of The Matrix, a person's perception of events in time does not speed up when danger looms. However, the volunteers *did* have a distorted view of time during their fall. Before they ascended the tower, Stetson asked each volunteer to reproduce how long a compatriot took to hit the net using a stopwatch. They were then asked to do the same after they'd had a go themselves. On average, the volunteers estimated that own experience took 36% longer than that of their fellows. Time did not slow down – the volunteers just remembered that it did. Stetson and co believe that people lay down richer, denser memories when they experience shocking events. These 'flashbulb memories' include emotional content, which involves the brain's emotional centre – the amygdala. As they played back, their unusual richness could fool the brain into thinking that the recorded events took up more time than was actually the case.

Symbols, jugs, pizzas, balls

You all know the score. A train leaves one city travelling at 35 miles per hour and another races toward it at 25 miles an hour from a city 60 miles away. How long do they take to meet in the middle? Leaving aside the actual answer of four hours (factoring in signalling problems, leaves on the line and a pile-up outside Clapham Junction), these sorts of real-world scenarios are often used as teaching tools to make dreary maths "come alive" in the classroom. Except they do not really work. A new study shows that far from easily grasping mathematical concepts, students who are fed a diet of real-world problems fail to apply their knowledge to new situations. Instead, and against all expectations, they were much more likely to transfer their skills if they were taught with abstract rules and symbols.

The use of concrete, real-world examples is a deeply ingrained part of the maths classroom. Its advantages have never really been tested properly, for they appear to be straightforward. Maths is difficult because it is a largely abstract field and is both hard to learn and to apply in new situations. The solution seems obvious: present students with many familiar examples that illustrate the concepts in question and they can make connections between their existing knowledge and the more difficult concepts they are trying to pick up. The train problem is a classic example. Another is the teaching of probability with rolls of a die, or by asking people to pick red marbles from a bag containing both blue and red ones. The idea is that, armed with these examples, students will recognise similar problems and apply what they have learned. It is a technique deeply rooted in common sense, which is probably as good an indicator as any that it might be totally wrong.

Jennifer Kaminski from Ohio State University demonstrated this by recruiting 80 undergraduate students and teaching them about a simple mathematical concept that involved adding three separate elements together.[72] The concept included very basic mathematical ideas, like the concept of zero, or the idea of commutativity – that the order in which things are added does not change the result (1+2 and 2+1 both equal 3). Three groups were taught using familiar, concrete examples. The first group were told to imagine measuring cups containing varying levels of liquid and asked to work out how much liquid remained when the two were combined. So for example, combining a jug that was a third full with another that was two-thirds full would give a full jug. Uniting two jugs both two-thirds full would give one jug that was a third full as a remainder. The second group was taught using another similar example involving pizza slices alongside the jugs, and the third group learned both of these, along with a third system involving tennis balls. They were told that each new system worked in the same ways as the old ones, and obeyed similar rules. The fourth group was taught in a more generic way. The vivid jugs, pizzas and balls were replaced with a meaningless and arbitrary symbols, and the students simply had to learn how they could be combined. For example, a circle and a diamond combined to form a wavy rectangle, while the sum of two circles was a diamond.

After the training, feedback and 24-question multiple-choice tests, Kaminski was satisfied that the vast majority of students had picked up the principles successfully. She then asked to students to apply their knowledge to a fresh setting, described as a foreign children's game involving three objects. Children pointed to two of the objects and one child, who was "it", had to point to the correct final one to win. The students were told that the game's rules were very much like those of the systems they had just learned. They were shown some examples so that they could deduce these rules and were tested with 24 multiple-choice questions. These were, in fact, the same as the questions they had previously answered, but "translated" to the new setting. Contrary to all expectations, the group who were taught with generic symbols fared best, answering 76% of the questions correctly. They completely outperformed the three groups who were taught with real-world examples, who all scored between 44-51%, no better than a chance result. The group that was taught using three concrete examples fared just as poorly as than the one which only learned one system. So the common sense idea that students are capable of picking out the similar threads from related examples seems to fall short. In fact, quite the opposite was true – they were better able to apply their knowledge to a new setting if they were taught using generic abstract examples.

The result seems so at odds with the traditional view of education that Kaminski tested it further. She recruited another 20 volunteers and taught them the jug and pizza systems, but this time, she explicitly spelled out the similarities between the two. Astonishingly, this did not help matters; the students' scores still reflected random guesswork more than learned problem-solving. If the students were asked to work out the similarities between the two systems themselves, about half of them scored very high marks of 95%, but the others still performed no better than chance. So this type of teaching method helps some high-performing students to achieve a top grade, but it also fails to benefit others. And on average, the 'class' still scored less than the group who learned the generic symbols.

In a final experiment, Kaminski wanted to see if a combination of real-world and generic examples would have a stronger effect than either alone. Clearly, they both have their advantages – generic examples seem easier to apply, while real-world ones are easier to pick up at the start. Even so, students who only learned the abstract symbol system *still* outperformed a second group who learned the jugs method, followed by the symbols.

Kaminski's work will no doubt come as a shock to those in the education sector. While it is certainly true that students engage with mathematical concepts more easily when faced with real-world examples, these striking experiments suggest that they may not actually be picking up any real insights about the underlying principles. And without those, they are unable to apply their knowledge from one real-world example to another, exactly the opposite of what maths teachers want to achieve! Kaminski is not calling for an end to all real-world examples in classrooms, but she suggests that they should only be used when the basic abstract principles have been introduced. Deeply grounding an abstract concept in a real-world example could actually do more harm than good, by constraining the knowledge that students gain and hindering their ability to recognise the same concept elsewhere. Questions about bags of marbles and speeds of trains really are just about bags of marbles and speeds of trains.

Social exclusion literally feels cold

Our languages are replete with phrases that unite words evoking a sense of cold with concepts of loneliness, social exclusion or misanthropy. When we speak of icy stares, frosty receptions and cold shoulders, we summon up feelings of isolation and unfriendliness. But cold and solitude are more than just metaphorical bedfellows; a set of cunning psychological experiments show that social exclusion can literally make people feel cold. Chen-Bo Zhong and Geoffrey Leonardelli from the University of Toronto recruited 65 students and asked them to recall a situation where they either felt included within a group or left out of it.[73] Afterwards, they asked the students to estimate the temperature in the room under the ruse of providing information for the maintenance staff. The estimates varied wildly but volunteers who had social exclusion on their minds gave an average estimate of 21°C, while those who remembered fitting in guessed an average of 24°C.

Zhong and Leonardelli were not content with simply bringing social memories to the surface; for their next trick, they created real feelings of exclusion. They recruited 52 more students who were led to a computer cubicle and told that they were taking part in an online game with three anonymous players. The object was simply to throw a virtual ball between the group but unbeknownst to the volunteers, their "partners" were computer-controlled programmes. These avatars either intermittently passed the ball to the real player throughout the game, or left them out after a few cursory passes. Afterwards, the students completed an apparently unrelated marketing survey where they had to rate their preference for five different foods or drinks. Zhong and Leonardelli found that the behaviour of their virtual peers affected their preferences for foods, depending on their temperature! The "unpopular" students who had been left out of the ball-throwing game showed significantly stronger preferences for hot coffee or hot soup than those who had been allowed to play. On the other hand, both groups showed the same degree of preference for control foods such as Coke, apple or crackers.

So being ostracised, or even the *memory* of being ostracised, drums up both a literal chill and a desire for warmth. It can work the other way round too; invoking the concept of temperature can alter your opinions of another person. In another study, Lawrence Williams at the University of Colorado and John Bargh from Yale University shows that warming a person's fingertips can also bring out the warmth in their social relationships, nudging them towards judging others more positively and promoting their charitable side.[74] The duo recruited 41 volunteers and when they arrived at the psychology building, a colleague (who was not aware of the experiment's goals) escorted them to the laboratory and asked them to hold a cup of coffee for her along the way. Once in the lab, they had to read a description of a stranger and rate them on 10 different personality traits. The cups of coffee were the key element. Half were hot and half were iced, and the volunteers' brief contact with the cups was enough to sway their later impressions. The recruits whose hands were heated by their cups rated the stranger as having a warmer personality than those who held the cold cups. On average, they gave him a score of 4.7 on a scale of 1 to 7, while the cool-handed people gave average scores of 4.3. The difference was small but statistically significant, and it was not just a symptom of a generally improved mood brought on by a steaming cup of java. After all, Williams and Bargh also found that the temperature of the coffee cups had no bearing on how the recruits judged the stranger along personality traits unrelated to warmth of personality. There was, however, a small chance that the accomplice was subtly influencing the volunteers' behaviour since she too had clutched the cups of coffee.

With that in mind, Williams and Bargh gave another group of 53 people an envelope of instructions, asking them to hold a hot or cold therapeutic pad under the premise of evaluating it. Later, when the unknowing volunteers were offered a small reward for their trouble – a bottle of Snapple or a dollar voucher for the local ice-cream parlour – those who touched the hot pads were more likely to give it to a friend than to keep it for themselves. Three-quarters of them chose this charitable option, compared to just 54% of the recruits who held the cold pad. Together, the coffee and pad experiments show that sensations of can both affect a person's judgment of, and influence their actions towards, other people. In both cases, the recruits made their choices freely and were unaware that they had been subtly manipulated.

The small size of the effect might make it easy to dismiss the result as a kooky psychological curio; after all, in a real-life situation, surely other factors like sense of humour or personal hygiene would have a greater impact on first impressions? Perhaps so, but psychological research has shown that the warm-cold dimension is incredibly important to us when we meet other people. Our first impressions of strangers tend to be based primarily on how warm we think they are and, to a lesser extent, how competent. We make warm-cold judgments quickly and automatically, and they provide us with a quick overview of other important personality traits including friendliness, trustworthiness and helpfulness.

Studies like these are a testament to the power of metaphors. When we first learn to wield them in English lessons, we are taught that they are not meant to be taken literally, and yet psychological experiments show that many metaphors reflect fundamental ties between our social lives and our physical sensations. This close link between temperature and social closeness may be rooted in our earliest interactions with other people. When our parents held us close to them, we felt the warmth of their bodies and when they kept their distance, they deprived us of that warmth. From an early age, temperature and distance from another person go hand-in-hand. Later on in life, the heat of a cup of coffee may recall these earlier warm experiences and the feelings of trust and comfort that are associated with them.

Indeed, people seem to have a tendency to describe complex abstract concepts using familiar physical experiences. Positive traits like generosity, friendliness or compassion are associated with warmth, while greed and desire are associated with hunger. So it is with cold and solitude, warmth and friendliness. This link between the physical and the abstract is not just a one-way street - the two domains are so closely linked that the abstract can also invoke the physical. The results open a door to a large flurry of follow-up questions. Could experiencing physical warmth help to lessen the hurt of social exclusion or improve the chances of making new friends? Could a cup of warm chicken soup be more than a metaphorical remedy "for the soul"? Could ambient temperate affect the quality of social relationships, and could manipulating one influence the other? Does social cohesion have its own thermostat? Could the depressive experience of seasonal affective disorder (SAD) stem just as much from the chill of winter as it does from a dearth of sunlight? Are fans of Vanilla Ice destined to a life of loneliness and despair?

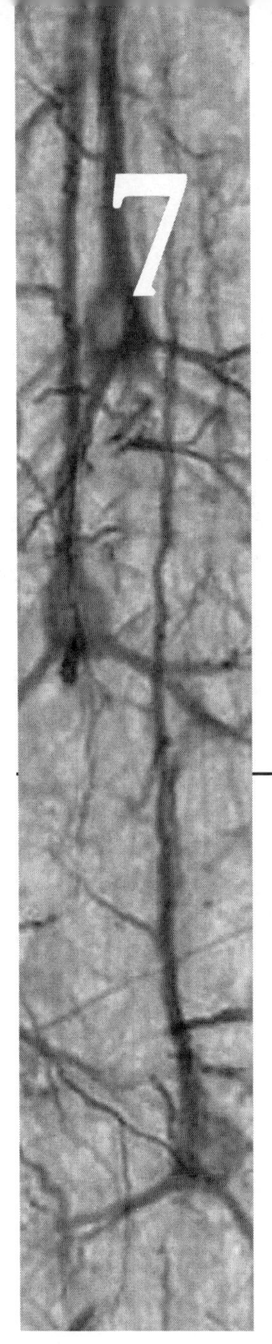

Nervous science

The neuroscience of optimism,
art appreciation and
jazz improvisation

Your brain on improvisational jazz

You're lying face-up with your head surrounded by a large medical scanner. You have been told that you have to keep your head completely still. You cannot look down to see your hands and you cannot hear very much beyond the background hum of the machine. A perfect time for some improvisational jazz then....

This was the challenge posed to six professional jazz musicians by Charles Limb and Allen Braun from the National Institute on Deafness and Other Communication Disorders. They were told to play that funky music while undergoing fMRI (functional magnetic resonance imaging), a brain-scanning technique that would enable the two researchers to quite literally see the creative process in action.[75] Improvisation is the cornerstone of great jazz performances, and it is a joy to watch musicians deftly adapting pieces of music on the fly to create something familiar yet wholly original. To Limb and Braun, this skill is a fine example of the creative process, itself a "quintessential feature of human behaviour". The duo approached their study with the assumption that there is nothing mystical or obscure about creativity. Like all other matters of the mind, it is the result of firing neurons – a tangible process that we can understand and even see in action.

That is where the fMRI came in, but getting someone to play properly within such restrictive confines is challenging to say the least. To do it, Limb and Braun ordered a special, custom-built piano keyboard, which lacked any metallic parts that would be affected by the scanner's powerful magnets. Each musician lay down in the scanner with their knees up and the keyboard in their laps. An angled mirror positioned over their eyes allowed them to see the keys, and a small earpiece allowed them to hear what they were playing. In the 'Scale' task, the musicians were either asked to repeatedly play the simple C major scale within one octave, or to play an improvised melody of their choice using only the same nine notes. This was a fairly basic form of improvisation and Limb and Braun also wanted to capture the much richer nature of true jazz performances. In the more complicated 'Jazz' task, they were first asked to learn an original jazz composition. Once in the scanner, they either had to play the unaltered tune, or to improvise freely, using the chord structure of the new piece as a guide. In both cases, they listened to a pre-recorded accompanying quartet, which they could use as inspiration during the improvisational phase.

The musicians showed very similar patterns of brain activity during both the Scale and Jazz experiments, despite the different degrees of difficulty. In both cases, certain parts of the brain were consistently activated during improvisation, while others were consistently turned off. The prefrontal

cortex, which controls many of our higher mental abilities from planning to problem-solving, was particularly strongly affected. Large swathes of it were completely shut down, including the LOFC (lateral orbifrontal cortex), which monitors and blocks out inappropriate behaviours, and the DLPFC (dorsolateral prefrontal cortex), which is involved in planning and focused, methodical thinking. These widespread deactivations could help musicians to avoid over-thinking a task. By turning off parts of the brain involved in self-assessment and focused attention, they could pave the way for actions and ideas that are spontaneous, unplanned and new.

Within the large prefrontal cortex, only one area was strongly *activated* during improvisation – the medial prefrontal cortex (MPFC). Recent studies have suggested that the MPFC plays a role in our sense of self, and it certainly lights up when people recount personal anecdotes. To Limb and Braun, its heightened activity fits with the idea that musicians use improvisation as an outlet for expressing their own musical voice and experiences. The fMRI images also revealed that when improvising, the musicians' brains showed more activity in areas involved in hearing, sight and touch, even though they were seeing, touching and hearing much the same things as they were when playing from memory. Their improvised pieces were no faster than the originals nor did they include more notes. Instead, Limb and Braun suggest that during the creative state, a musician's senses are heightened across the board.

The scans imply that there is a very consistent pattern of brain activity linked to creativity, a pattern of heightened senses and self-expression with a lack of conscious control. And the very similar results from the Scale and Jazz scans suggest that this has nothing to do with the complexity of the music and everything to do with the creativity of the performers. Limb and Braun also tantalisingly suggest that a similar prefrontal shut-down could also be linked to other states of altered consciousness including meditation, hypnosis and most intriguingly, dreaming.

Brain-training

Forget 'smart drugs' or brain-training video games. According to new research, a deceptively simple memory task can do what no drug or game has done before – it can boost your 'fluid intelligence', your ability to adapt your powers of reasoning to new challenges. Fluid intelligence does not rely on previous knowledge, skills or experience. It is at work when we solve new problems or puzzles, when we draw inferences and spot patterns, and when we test ideas and design experiments. Fluid intelligence appears to be strongly influenced by inherited genetic factors and is closely related to success in both the classroom and the workplace. The ability plays such a central role in our lives that it begs an obvious question: is there any way of training your fluid intelligence?

Video game manufacturers would like you to think so. Games like Dr Kawashima's Brain Training and Big Brain Academy are suggestively marketed as ways of improving your brain's abilities through the medium of number problems, Sudoku and word puzzles. As a result, your brain will allegedly become younger. The pitch is certainly a successful one – these games are bestsellers and are increasingly copied by a swarm of imitators. Last year, the worth of the US brain-training market alone was estimated at about $80 million. Whether these products actually work is open to debate but there is certainly no strong evidence that they do anything beyond improving performance at specific tasks. That seems fairly obvious – people who repeatedly practice the same types of tests, such as number sequences, will become better at them over time but may not improve in other areas, like memory or spatial awareness. But acquiring advanced skills in one specific task is a far cry from increasing your overall fluid intelligence; it would be like saying that you are a better musician because your scales are second-to-none.

Nonetheless, Susanne Jaeggi from the University of Michigan has developed a training programme involving a challenging memory task, which appears to improve *overall* fluid intelligence.[76] The trainees do better in intelligence tests that have nothing to do with the training task itself and the more training they receive, the higher their scores. Jaeggi's work was inspired by Graeme Halford from the University of Queensland, who suggested that the limits of our reasoning abilities are very similar to the limits of our working memory. The term "working memory" refers to our ability to temporarily hold and manipulate pieces of information, as you do when you add up prices on a bill. Reasoning and working memory are not identical, but they both involve holding pieces of information in a sort of mental notepad, and even seem to rely on similar networks of neurons. The idea is that both are constrained by the processing power of our brains and our ability to focus our attention. So Jaeggi reasoned that a task that improved working memory would also give fluid intelligence a boost.

She recruited 70 young students and set half of them on a challenging training regime, involving the so-called "n-back task". These trainees watched a series of screens where a white square appeared in various positions on a black background. Each screen appeared for half a second, with a 2.5 second gap before the next one flashed up. While this happened, the trainees also heard a series of letters that were read out at the same rate. At first, their job was to say if either the screen or the letter matched those that popped up two cycles ago but the number of target cycles increased or decreased depending on how good the students were at the task. Boffins eventually had to compare the current pair with those many cycles ago, while dunces only had to remember fairly recent ones. The task is challenging because students had to continuously hold a large amount of regularly updated information in their heads – a goal that taxed their working memory to the limit.. The students sat through about half an hour of training a day for either 8, 12, 17 or 19 days, and were tested on their fluid intelligence before and after the regimen using the standard German Bochumer-Matrizen Test.

On the whole, the trainees did significantly better on the fluid intelligence test than their peers who did not receive any training. The control group did improve slightly, as you might expect from people who had done similar tests in the past, but the trainees still outperformed them. And their degree of improvement depended on the extent of their training. Those who were trained for the longest time – 19 days – showed the biggest improvements, while those who were only trained for eight days did not get significantly better. Jaeggi noted that the test did not just hone the abilities of the naturally intelligent people in the group, as low-performers benefited from the training programme just as much as high-performers did, if not more so. Nor did those who already had powerful working memories enjoy greater benefits, which suggests that the training does not simply work by improving this specific skill. Jaeggi thinks that this task worked where others have failed because it remained challenging. The students were never allowed to get comfortable with the task – as soon as they improved, it became accordingly more difficult. Faced with the combination of two info streams and shifting difficulty levels, they could not develop simple strategies or switch to autopilot. Even without this tweak, the task was fundamentally very challenging. To succeed in it, students had to remember old items, constantly update the memories they were keeping, block out irrelevant ones, and manage two tasks at the same time using both their sense of sight and hearing.

The results seem promising and the prospect of a single training exercise that can improve a whole suite of mental abilities certainly seems like a good thing. But for now, the study leaves behind many unanswered questions. How exactly does the training programme lead to better fluid intelligence? At what point will the benefits of extra training start to level off? And how long will it take for the programme's effects to wear off, it they ever do? The answers to these questions will help to decide if the findings are indeed "highly relevant to applications in education" as the authors claim.

Beauty in the brain of the beholder

Is beauty simply in the eye of the beholder, or do all the beholders' brains have something in common? Is there an objective side to beauty? Plato certainly seemed to think so. His view was that beauty was an inherent property that all beautiful objects possess, irrespective of whether someone likes them or not. To him, beauty in the world stemmed from an ideal version of Beauty that real objects can only aspire to. A biologist might instead suggest that the objective side of beauty stems from built-in predispositions for certain features, colours, shapes or proportions. The opposing view is that art is a fully subjective enterprise and our preferences are shaped by our values and experiences. The real answer is likely to lie somewhere in the middle – after all, art students learn basic common skills such as proportion, perspective and symmetry before embarking on their own stylistic journeys.

Artists, critics and gallery visitors can argue about this question all they like, but some clearer answers have now emerged from three researchers in Italy, arguably the home of the some of the world's most beautiful art. Cinzia Di Dio, Emiliano Macaluso and Giacomo Rizzolatti from the University of Parma have brought the tools of the modern neuroscientist into the debate.[77] They showed images of 15 masterpiece sculptures from the Classical and Renaissance periods to 14 regular people who had no experience of art criticism. As they looked at the pictures, the scientists scanned their brains with functional magnetic resonance imaging (fMRI). Some of the images were subtly altered. In their original form, the sculptures were proportioned according to the Golden Ratio, the ratio of lengths that the ancient Greeks thought was most aesthetically pleasing. It works out to about 1.16 and turns up all over the place in classical art and architecture. For example, if you take the sculpture *Doryphoros* by Polykleitos and divide the distance from his navel to his knee by the distance from his navel to his foot, you get 1.16. The ratio of the distances from navel to sole and from navel to head is also 1.16. In the altered images, the statues had different proportions but were the same in every other respect. Their trunks or legs were stretched enough to throw their golden proportions out of kilter, but not enough for the viewers to consciously twig that something was wrong.

At first, the researchers simply asked their subjects to look at the images as if they were in a museum, without having to perform any complicated task. On average, they gave the original images positive reviews 76% of the time, but only liked the distorted images 63% of the time. The golden ratios of the original images strongly activated neurons in a part of the brain called the insula, while the altered images did not. The insula is involved in the feeling of emotions and the research trio speculate that it generates the positive feelings that accompany beautiful images. To them, these results show that beauty is at least partially objective. They suggest that groups of neurons in the brain are trained to respond to different characteristics, such as proportion, and have certain preferences. When the signals from these individual processing centres arrive at the insula together, it generates a positive feeling – a sense of beauty.

But the team also showed that subjectivity still exerts a sizeable influence. They repeated their experiment but asked the viewers to judge the images, generally in terms of aesthetics and specifically based on their proportions. They compared their brain scans as they responded to images that were consistently praised for their beauty with those that were constantly denounced for their ugliness. This time, the insula did not matter. Instead, the beautiful images activated the amygdala, a region that links learned information with emotions.

Together, these results strongly argue that even an abstract concept like beauty is the product of the very tangible and physical matter of the brain. Finding beauty in art is the result of a neural agreement between two parts of the brain that respectively govern the subjective and objective sides. When you stare at a painting, scene or object, the neurons that report in to your insula set an objective standard for assessing beauty. However, those of the amygdala draw on your emotional experiences to add a subjective colour to your final reaction. Di Dio, Macaluso and Rizzolatti end their paper by speculating that the objective side, driven by our biology, may influence the lasting quality of certain pieces of art. New trends arise because they appeal to either biological preferences or simply to novelty or fashion. But as the latter eventually fades and wanes, future generations may have only the former to base their judgments on.

The thought-controlled prosthetic arm

The realm of science-fiction has taken a big stride towards the world of science fact, with the creation of a prosthetic arm that can be moved solely by thought. Two monkeys, using only electrodes implanted in their brains, were able to feed themselves with the robotic arm complete with working joints. Bionic limbs have been fitted to people before but they have always worked by connecting to the nerve endings in the chest. This is the first time that a prosthetic has been placed under direct control of the relevant part of the brain. The study, carried out by Meel Velliste from the University of Pittsburgh, is a massive leap forward in technology that lets the brain interface directly with machines.[78] Previously, the best that people could do was to use a cap of electrodes to control the movement of an animated computer cursor in three-dimensional space. Velliste's work shows that signals from the brain can be used to operate a realistically jointed arm that properly interacts with objects in real-time. The applications of the technology are both significant and obvious. Amputees and paralysed people could be kitted out with realistic prosthetics that afford them the same level of control as their lost limbs once did.

The fake arm is certainly a good step towards realism, and can rotate freely about the shoulder and flex about the elbow. Only the hand is simplified; the 'wrist' is fused and the 'hand' consists of two opposing 'fingers'. Despite this claw-like tip, the monkeys learned to use the arm successfully and there were even signs that they came to accept it as their own. Velliste implanted an array of 116 tiny electrodes into the part of each monkey's brain that controls movement – the motor cortex. A computer programme analysed the readouts from these electrodes in real-time and converted them into signals that moved the arm. All the while, the animals own arms were restrained, forcing them to rely on their surrogate limbs.

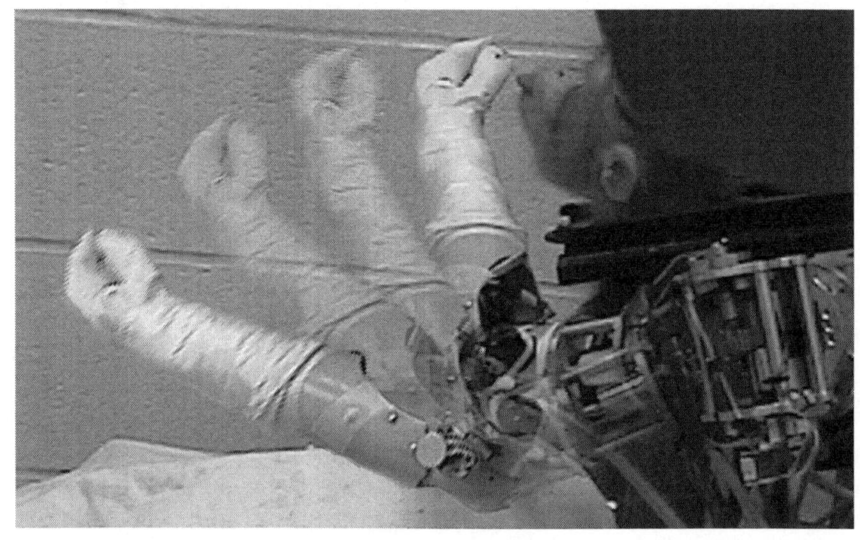

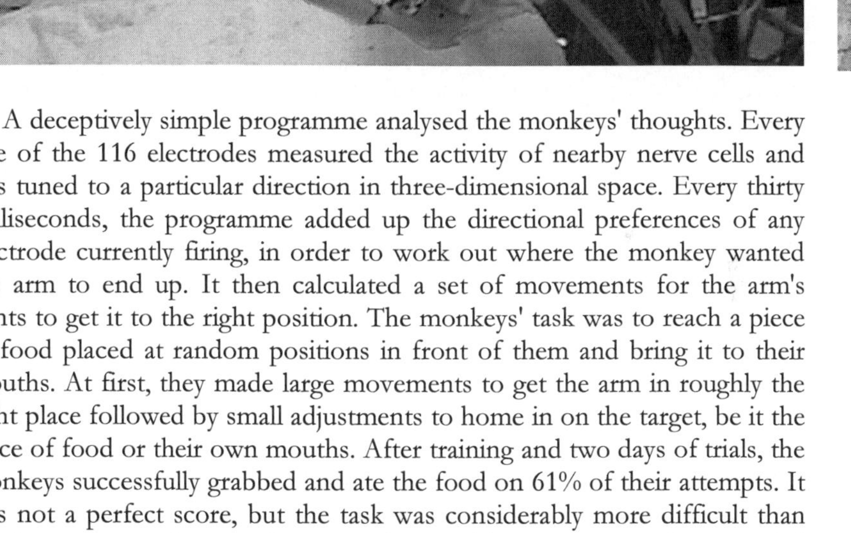

A deceptively simple programme analysed the monkeys' thoughts. Every one of the 116 electrodes measured the activity of nearby nerve cells and was tuned to a particular direction in three-dimensional space. Every thirty milliseconds, the programme added up the directional preferences of any electrode currently firing, in order to work out where the monkey wanted the arm to end up. It then calculated a set of movements for the arm's joints to get it to the right position. The monkeys' task was to reach a piece of food placed at random positions in front of them and bring it to their mouths. At first, they made large movements to get the arm in roughly the right place followed by small adjustments to home in on the target, be it the piece of food or their own mouths. After training and two days of trials, the monkeys successfully grabbed and ate the food on 61% of their attempts. It was not a perfect score, but the task was considerably more difficult than anything attempted in previous experiments. The same research group, led by Andrew Schwartz, had thus far only managed to get subjects to move a virtual cursor towards a target 80% of the time. In this study, the monkey succeeded in the equivalent challenge – moving the arm in the general direction of the food – 98% of the time. Beyond that, it also had to home in on the prize, grasp it and bring it back to its mouth.

Velliste believes that the secret to the arm's success lay in making it as natural as possible. For a start, the monkey could freely move its head and eyes without affecting the signals controlling the arm. It could also move the fake arm in real time. There was only about a seventh of a second worth of delay between a burst of brain activity and the corresponding movement; natural arms have similar delays between thought and deed. This nigh-instantaneous control was obvious during one trial when the animal dropped the food and immediately stopped moving the arm. This natural responsiveness made it easier for the monkeys to accept the arm as their own. They learned behaviours that had nothing to do with the task, like licking remaining food off the fingers or using the hand to push food into their mouths. They learned to move the arm in arcs to avoid knocking the food off the platform while bringing it back in a straight line. They even learned that the food (marshmallows and grape halves) stick to the fingers so while they initially opened the hand only when it was near their mouths, one of them figured out that they could open the hand well before then.

Behaviour like this is very promising, for it suggests that interfaces between brains and computers could be used to drive prosthetics in real-world situations. So far, the main flaw with the fake arm is that it is not quite as fast as a real one yet. The monkeys took about 3-5 seconds to complete the task, which is about twice as much time as they would take with their own arms. This may be due to the need for small corrective movements once the arm was in roughly the right place. The researchers also saw that one of the monkeys made small movements with its restrained right hand. Obviously, amputees would not have that luxury, but Velliste argues that it is unlikely that these movements made controlling the arm any easier. For a start, the electrodes were only implanted in the right half of the monkey's brain, which controls the left arm not the right. And we know from other studies that you do not need to move in order to control a virtual prosthetic.

Broken chains and faulty mirrors

Your brain has an amazing ability to predict the future. For example, if you see someone reach for a chocolate, you can guess that they are likely to pick it up, put it in their mouths and eat it. Like most people, you have a talent for understanding the goal of an action while you see it being performed – in this case, you know that reaching for the chocolate is only a step towards eating it. That may not sound very impressive, but as with many mental skills, it is only apparent how complicated it is when you see people who cannot do it. Autistic people, for example, find it incredibly difficult to relate to other people and this may, in part, be because they cannot understand the why of someone else's actions. While a typical child would understand that a mother holding her hands out is readying for a hug, an autistic child might be baffled by the gesture. Now, a new study by Luigi Cattaneo, Giacomo Rizzolatti and colleagues suggests that autistic people find it difficult to understand the purpose of an act because they cannot string together different actions into a coherent whole.[79] And underlying this problem is a special group of nerve cells called mirror neurons.

Giacomo Rizzolatti discovered mirror neurons about a decade ago and they have been arguably one of the most exciting finds in recent neuroscience. These cells fire not only when you perform an action but when you see someone else doing the same – a case of human-see, human-do. Scientists have suggested before that faults in the mirror neuron network are linked to autism. Some have found that autistic children have less mirror neuron activity than average, and parts of the brain where these cells are normally found tend to be thinner in autistics. Faults in the mirror network might render autistics unable to respond correctly to the actions of others, but Cattaneo and Rizzolatti have a different idea. They focused on a special group of mirror neurons in the parietal lobe that take in the bigger picture. These cells fire when we see a particular action, but they only fire strongly in certain contexts. One of Rizzolatti's earlier studies found that some mirror neurons fire when someone grasps an object, but some of these only fired strongly when they then put the object in their mouths rather than in a box. These special cells do not just tell their owner what someone else is doing, they provide a clue about *why* they are doing it. They divine intention, and that is exactly the type of skill that autistic people are bad at.

To test their ability to understand intentions, Cattaneo asked seven high-functioning autistic children and eight typical children aged 5-9 to watch an experimenter perform two simple actions. They either grabbed a piece of food and ate it, or grabbed a piece of paper and put it into a box strapped to the experimenter's right shoulder. As they watched, he measured the electrical activity in the mylohyoid muscle that opens the mouth. When the typical children saw the experimenter grab for the apple, this muscle became more active but grabbing the paper triggered no such response. Even when the experimenter was just *reaching* for the apple, the children were already aware of his final goal and their mouth-opening muscle tensed in hungry anticipation. But this did not happen in the autistic children – their mylohyoid muscles did not tense in either the apple or the paper trial. Cattaneo repeated the experiment again but this time, he asked the children to perform the action themselves. As expected, the mylohyoid muscles of the typical children became slightly active as they reached for the food, and reached its peak as their mouths started to open. Again, the autistic children behaved differently – their mylohyoids only became active when they brought the food to their mouths. Neither reaching for the apple, nor grabbing it, caused a twitch in the muscle.

To prove that this effect was not just to do with food and mouths, the team did another experiment where they asked the children to put a piece of food into a bin, which had to be opened with a foot pedal. This time, Cattaneo and Rizzolatti measured the activity of the tibialis anterior muscle that flexes the foot at the ankle. When the children pressed the pedal, the muscle relaxed, and as before, typical children started relaxing the muscle when they reached for the food, but autistic children only did so when they were actually bringing the apple to the bin.

When we plan an action, our brain chains together a series of motor commands and anticipates the endpoint before we begin moving. Even simple actions like "eat an apple" can be chains consisting of links like "reach for apple", "grab apple" and "bring apple to mouth". The fact that a child's mouth muscles start to tense in the first phase suggests that the brain knows what it is doing well in advance. According to Cattaneo and Rizzolatti, autistic children cannot string together these chains. Without the ability to foresee the goal of an action when it starts, an autistic child can work out what someone is doing, but is much hazier about *why* they are doing it. Rizzolatti is quick to point out that this does not mean that autistic children are at a complete loss. They could, for example, infer a person's motivations from context or habits. But they will always have a natural disadvantage because they cannot understand someone else's motives as if they were their own. This dysfunction could be the key to the large social problems that these children face.

The ups and downs of brain-enhancers

Most of my writing gets done late at night when tiredness starts creeping in, and a boost in concentration and attentiveness are sorely needed. In lieu of actually getting some sleep, the ability to pop a little pill that will have the same effect sounds pretty enticing. Unfortunately (or perhaps luckily), the closest thing I ever have available is the coffee in the kitchen. But for many people, taking a pill to sharpen their mental faculties – a so-called "cognitive enhancer" – is a much easier deal. A large number of prescription drugs can indeed give you a little mental boost, including amphetamine and methylphenidate (more familiarly known by its brand name Ritalin). Both the use and the range of such drugs are on the rise and they seem capable of stimulating debate just as readily as they do

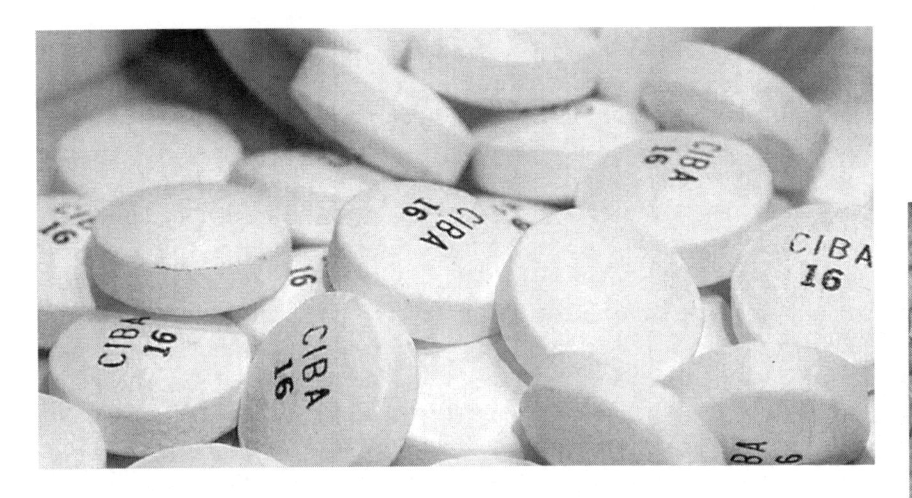

the brain. They have their medical uses; as Ritalin, methylphenidate is used to treat attention deficit hyperactivity disorder (ADHD) and, less commonly, narcolepsy. Even this is not without controversy, but the fact that they seem to have the same enhancing effects in healthy people opens up the potential for recreational use. And *that* is far more divisive.

Last year, the journal *Nature* published a commentary which looked into the ethics of such drugs and sparked off a heated online debate in their Nature Network forum. More recently, the magazine released the results of an informal survey of over 1,400 readers, which showed that about 20% admitted to using cognitive enhancers for non-medical reasons and a far higher proportion approved of such use. The ethical issues at stake are incredibly broad, but one specific problem is that cognitive enhancers represent a case of technology outpacing science. Common though these drugs are, we still do not fully understand how some of them work. Take methylphenidate. At a basic level, we know that it interferes with protein pumps that import two of the brain's signalling molecules – dopamine and norepinephrine – into neurons. As a result, these molecules build up in the spaces between neurons – the synapses. But why should such a build-up improve a person's performance?

Nora Volkow, Director of the National Institute on Drug Abuse, thinks she has the answer. By studying the metabolic activity of brains dosed up with methylphenidate, she has found evidence that the drug works by focusing the brain's activity and making it more efficient.[80] And crucially, the benefits (and costs) you reap from that may depend on how focused your brain already is. As our brains engage in any mental task, sets of neurons start to fire but this extra activity takes place among a lot of background neural noise, like a person shouting in a crowded bar. Earlier studies have suggested that dopamine and norepinephrine lower the levels of background buzz so that any extra activity directly related to the task at hand stands out. In our metaphor, the chemicals silence the noisy bar. In engineering language, they increase the brain's 'signal-to-noise ratio'. To test this idea, Volkow used a PET-scanner to measure the amount of glucose (sugar) used by the brains of 23 volunteers as they did some mathematical calculations. The volunteers were injected with either methylphenidate or a placebo and asked to solve simple arithmetic problems, whose difficulty had been tailored to their individual abilities. As a control, they were asked to look at, but not respond to, images of scenery.

When faced with the pretty pictures, the volunteers brains behaved in the same way regardless of what they were injected with. It was only when they had to do the more complex mental task that the effects of methylphenidate were revealed. The brains of both groups burned up more glucose but people who had been shot up with placebo used up 21% extra sugar while those that were drugged with methylphenidate only needed half as much – an extra 11%. These figures suggest that the drug is indeed focusing the brain's activity and dramatically reducing its energy demands. The PET-scans said the same thing, for when the placebo volunteers did their sums, the area of extra activity in their brains was three times greater than the same area in the drugged volunteers. With this cerebral focusing, a brain on methylphenidate becomes more efficient than a normal one. Methylphenidate not only reduced the total brain activity in the calculating subjects, it also reduced the *number of brain regions* stimulated by the task. The silenced areas included the "default network", a collection of regions that is most active when our brains are resting and our minds are wandering, and is deactivated when the brain has tasks to do. Volkow speculates that methylphenidate boosts the brain's performance by helping this process along, corralling a wandering mind and allowing the brain to focus on the job at hand.

But not everyone reacts in the same way. Volkow found that the effect of the brain-boosting drug depended on the original state of the brain it boosted. Methylphenidate had the least impact on brains that had naturally high metabolic rates and burned up the least extra glucose when doing sums. She suggests that brains that are *already* efficient and well-focused do not benefit from cognitive enhancers. In fact, if people had already deactivated their default networks to the ideal extent, throwing methylphenidate into the fray may even worsen performance, as was the case in four of the volunteers. Volkow's work could help to explain why cognitive enhancers like methylphenidate can hone brain performance in some people and some contexts, but be equally detrimental in others. Sleep-deprived individuals, or those with ADHD, may benefit from a chemical that deploys their brain's resources in a more efficient way. People whose brains are already working at their best could actually suffer from being focused any further.

Decisions please, hold the consciousness

Our brains are shaping our decisions long before we become consciously aware of them. That is the conclusion of a remarkable study which shows that patterns of activity in certain parts of our brain can predict the outcome of a decision seconds before we're even aware that we're making one.

It seems natural to think that we carry out actions after consciously deciding to do so. I decide to start typing and as a result, my hands move around a keyboard. But according to modern neuroscience, that feeling of free will may be an illusion. For over twenty years, experiments have suggested that, unbeknownst to us, a large amount of mental processing goes on unconsciously before we become aware of our intention to act. The first such study was done by Benjamin Libet in 1983. Libet asked volunteers to press a button at a time of their choosing, and to remember the position of the second hand of a wristwatch when they first felt the urge to move. While this happened, Libet measured the activity of their supplementary motor area (SMA), a part of the brain involved in planning movements. Astonishingly, he found that the SMA became active about half a second *before* the volunteer felt a conscious desire to press the button.

The seminal experiment suggested that the brain makes decisions on a subconscious level and people only believe that they consciously drive their actions in hindsight. The results seemed to put a dent in beliefs about free will and understandably, it remains quite controversial. Some have criticised the techniques that Libet used, claiming that inaccurate measurements could explain the small gap between brain activity and conscious awareness. Even for people who were convinced by the experiment's results, a slew of important questions remained. Is the SMA the source of the decision, or is it responding to other parts of the brain even higher up the chain of command? And is this unconscious activity a sign that the relevant areas of the brain are revving into readiness, or does it actually predict the action that is eventually taken? Now, Chong Siong Soon and colleagues from the Max Planck Institute have addressed these queries with an elegant update of Libet's work.[81]

Soon asked 36 volunteers to watch a letter in the centre of a computer screen, which changed every half-second. As the letters streamed by, the volunteers had to press one of two buttons, whenever they felt like it. When that happened, they were given a choice of four letters and had to choose the one that was up when they made their button-pressing decision. As they did this, Soon measured their brain activity using a scanning technique called functional magnetic resonance imaging (fMRI). He then used a statistical technique called pattern recognition to see if he could match the patterns of activity in different parts of the brain with the choice to press the right or left button. On average, Soon found that the volunteers took about 22 seconds to press the button and felt that they consciously decided to do so about a second or less before they made the movement. But the fMRI data told a much different story. Two parts of the brain – the frontopolar cortex and the precuneus – showed activity that predicted the choices that the volunteers made and in the frontopolar cortex, this activity happened a whopping *seven seconds* before the subjects were consciously aware of their decisions.

These astonishing results suggest that by the time we become consciously aware of a decision to move, our choices have already been influenced for several seconds by the actions of the frontopolar cortex. The study goes well beyond Libet's original work; it shows that this preliminary activity is far from a general and non-specific curiosity, but can actually predict a decision. Nor can it be explained away by inaccuracies in measurement – the timescales involved were far too long for that. These associations have never been seen until now because neither the frontopolar cortex not the precuneus were more active in *total* in the time leading up to the button press. Instead, it was the *pattern* of firing neurons within these areas that predicted the final decision, and it took the use of pattern recognition techniques for this effect come to light.

The involvement of the frontopolar cortex is not surprising. It fulfils the role of an executive manager and is involved in retrieving memories and controlling other high-level parts in the brain. Soon thinks that it is the source of the decision itself, with the precuneus simply storing the decision until it reaches a conscious level. When he changed his experiment so that volunteers were shown a cue to tell them when to make their choice, the frontopolar cortex still showed predictive activity before the signal, but the precuneus only did so in the time between signal and action.

Soon found, as Libet did, that the SMA was also active before the volunteers became conscious about their intentions. But he also showed that its pattern of activity can predict *when* the final decision is made. Again, this information is available at an unconscious level about five seconds before the volunteers actually move. The frontopolar and parietal cortex are not involved in timing until the few milliseconds before the movement, so it seems that the two parts of the brain have different and complementary duties. One shapes the outcome of a choice and the other affects its timing. Soon tentatively suggests that the sparks of a decision begin in the frontopolar cortex. From there, the decision is 'prepared' by the buildup of activity in the precuneus and later, the SMA. It is held there for a short while before we become consciously aware of it and act. This unconscious side to decision-making process may be for our own good. At least one experiment showed that people with damage to the relevant parts of the brain do not show any signs of unconscious preparation and make poorer decisions in a gambling experiment.

Studies like these have important philosophical implications. If our brains unconsciously make our decisions for us, is there any room for free will? Libet himself thinks so, but only in a restricted way. He asserts that for all the brain's unconscious preparation, people can still consciously decide to stop performing an action in the final milliseconds before thought becomes deed. In this view, it is more a case of "free won't" than free will.

ADHD: Delay not deviance

Attention-deficit hyperactivity disorder is the most common developmental disorder in children, affecting anywhere between 3-5% of the world's school-going population. As the name suggests, kids with ADHD are hyperactive and easily distracted; they are also forgetful and find it difficult to control their own impulses. While there is some evidence that ADHD brains develop in fundamentally different ways to typical ones, other results have argued that they are simply due to delay in the normal timetable for development. Philip Shaw, Judith Rapaport and others from the National Institute of Mental Health have found new evidence to support the second theory. According to their brain-scanning studies, ADHD is what happens when some parts of the brain stick to their normal timetable for development while others lag behind.[82]

The idea is not new; earlier studies have found that children with ADHD have similar brain activity to slightly younger children without the condition. Rapaport's own group had previously found that the brain's four lobes develop in very much the same way, regardless of whether children had ADHD or not. But looking at the size of entire lobes is a blunt measure that, at best, provides a rough overview. To get an sharper picture, they used magnetic resonance imaging (MRI) to measure the brains of 447 children of different ages, often at more than one point in time. At over 40,000 parts of the brain, they noted the thickness of the child's cerebral cortex – the brain's outer layer, where its most complex functions like memory, language and consciousness are thought to lie. Half of the children had ADHD and using these measurements, Shaw could work out how their cortex differed from typical children as they grew up.

As a child grows, their experiences manifest as connections between nerve cells, whose increasing number cause the cortex to thicken. But during adolescence, the developing brain values efficiency over expansion and the cortex starts to thin as unused connections are mercilessly trimmed. The growth of a child's brain into a teenager's is like the pouring of a block of clay that can then be sculpted away into the refined adult version. In both groups of children, with or without ADHD, parts of the cortex peaked in terms of thickness in the same order, with waves of maturity spreading from the edges to the centre. The pattern was the same, but the timing was not. On average, the brains of ADHD children matured about three years later than those of their peers. Half of their cortex had reached their maximum thickness at age 10½ , while those of children without ADHD did so at age 7½ . According to these results, ADHD is a disorder of delay, not deviance.

These delays were most pronounced in the lateral prefrontal cortex, where the lag time was as high as five years. These parts of the brain are responsible for suppressing inappropriate thoughts and actions, directing attention, short-term memory and controlling movement. All of these are tasks that children with ADHD can find difficult and other studies have found that as they try, their prefrontal cortex shows less activity than expected for a child of the same age. The only part of the brain that matured *faster* in children with ADHD was the primary motor cortex, which helps to plan and control movements. This area takes orders from the prefrontal cortex and if one matures early and the other matures late, this might explain several hallmarks of ADHD, including restlessness, fidgeting and uncontrolled hyperactivity.

Like much good research, Shaw's study raises more questions than it answers. For the moment, the most pressing one is: what causes the delay? From his data, Shaw rules out intelligence and gender, and thinks that prescribed drugs are unlikely to have an effect either. Genes, however, almost certainly have an influence and Shaw has his eye set on genes that produce a group of proteins called neurotrophins. These control the growth, division and survival of neurons, and changes in some of their genes have already been linked to ADHD.

Shaw's results should also be encouraging for many families, and they explain why so many children eventually grow out of the condition – as lagging brains catch up, the symptoms of the developmental lag might disappear. But it will be interesting to see if the timing of development at specific parts of the brain relates to a child's chances of recovery. We will only know that if scientists can run larger studies where the brains of children with ADHD are regularly scanned over a long period of time. Studying the speed at which a child's cortex matures could also shed light on other mental conditions. For example, autistic children show the opposite pattern to those with ADHD – their cortices mature much earlier than those of their peers. On the other hand, the cortices of children with exceptionally high IQ mature later, even though they thicken unusually quickly in early childhood.

What am I looking at?

Modern brain-scanning technology allows us to measure a person's brain activity on the fly and visualise the various parts of their brain as they switch on and off. But imagine being able to literally see what someone else is thinking – to be able to convert measurements of brain activity into actual images. It is a scene reminiscent of the 'operators' in *The Matrix*, but this technology may soon become a reality. Kendrick Kay and colleagues from the University of California, Berkeley have created a decoder that can accurately work out the one image from a large set that an observer is looking at, based solely on a scan of their brain activity.[83] The machine is still a while away from being a full-blown brain-reader. Rather than reconstructing what the onlooker is viewing from scratch, it can only select the most likely fit from a set of possible images. Even so, it is no small feat, especially since the set of possible pictures is both very large and completely new to the viewer. And while previous similar studies used very simple images like gratings, Kay's decoder has the ability to recognise actual photos.

To begin with, Kay calibrated the machine with two willing subjects – himself and research partner, Thomas Naselaris. The duo painstakingly worked their way through 1,750 photos while sitting in an fMRI scanner, a machine that measures blood flow in the brain to detect both active and inactive regions. The scanner focused on three sites (V1, V2 and V3) within the part of the brain that processes images – the visual cortex (right). The neurons in the visual cortex are all triggered by slightly different features in the things we see. All of them have different 'receptive fields'; that is to say that they respond to slightly different sections within a person's field of vision. Some are also tuned to specific orientations, such as horizontal or vertical objects, while others fire depending on 'spatial frequency', a measurement that roughly corresponds to how busy and detailed a part of a scene is. By measuring these responses with the fMRI scanner, Kay created a sophisticated model that could predict how each small section of the visual cortex would respond to different images.

Kay and Naselaris tested the model by looking at 120 brand-new images and once again, recording their brain activity throughout the experience. To account for 'noisy' variations in the fMRI scans, they averaged the readouts from 13 trials before feeding the results into the decoder. The duo then showed the 120 new images to the decoder itself, which used its model to predict the pattern of brain activity that each one would trigger. The programme paired up the closest matches for the predicted and actual brain patterns and guessed the order of the images that Kay and Naselnaris had looked at. It was incredibly successful, correctly identifying 92% of the images that Kay looked at and 72% of those viewed by Naselnaris. Obviously, using the average of 13 scans is a bit of a cheat, and if the machine were to ever decode brain patterns in real-time, it would need to do so based on a single trial. Fortunately, Kay found that this is entirely feasible, albeit with further tweaking. Even when he fed the decoder with fMRI data from a single trial, it still managed to correctly pick out 51% and 32% of the images respectively.

The decoder could even cope with much larger sets of images. When Kay repeated the experiment with a set of 1,000 pictures to choose from, the machine's success rate (using Kay's brain patterns) only fell from 92% to 82%. Based on these results, the decoder would still have a one in ten chance of picking the right image from a library of 100 billion images. That is over a hundred times greater than the number of pictures currently indexed by Google's image search. Obviously, the technology is still in its infancy – we are still a while away from a real-time readout of a person's thoughts and dreams based on the activity of their neurons. But these experiments are proof that such a device is possible and could be a reality sooner than we think.

Engaging with WALL-E

With their latest film WALL-E, Pixar Studios have struck cinematic gold again, with a protagonist who may be the cutest thing to have ever been committed to celluloid. Despite being a blocky chunk of computer-generated metal, it is amazing how real, emotive and characterful WALL-E can be. In fact, the film's second act introduces a entire swarm of intelligent, subservient robots, brimming with personality. Whether or not you buy into Pixar's particular vision of humanity's future, there is no denying that both robotics and artificial intelligence are becoming ever more advanced. Ever since Deep Blue trounced Garry Kasparov at chess in 1996, it is been almost inevitable that we will find ourselves interacting with increasingly intelligent robots. And that brings the study of artificial intelligence into the realm of psychologists as well as computer scientists.

Jianqiao Ge and Shihui Han from Peking University are two such psychologists and they are interested in the way our brains cope with artificial intelligence. Do we treat it as we would human intelligence, or is it processed differently? The duo used brain-scanning technology to answer this question, and found that there are indeed key differences. Watching human intelligence at work triggers parts of the brain that help us to understand someone else's perspective – areas that do not light up when we respond to artificial intelligence. Ge and Han recruited 28 Chinese students and made them watch a scene where a detective had to solve a logical puzzle.[84] The problem-solver was either a flesh-and-blood human or a silicon-and-wires computer (with a camera mounted on it). In either case, their task was the same – they were wearing a coloured hat and had to deduce whether it was red or blue. As clues, they were told how many hats of each colour there were in total and how many humans/computers had also been given hats. They could also see one of these peers, and the hat they were wearing. It is an interesting task, for both the human and the computer in this mini drama were given the same information and had to make the same logical deductions to get the right answer. The only difference was the tools at their disposal – the human used good, old-fashioned brain power while the computer relied on a program.

The students' job, as they watched this scene, was to work out if the problem-solver was capable of divining the colour of their hat. As the volunteers reasoned their way to an answer, Ge and Han scanned their brains using a technique called functional magnetic resonance imaging (fMRI). They found that the group who watched the humans showed greater activity in their precuneus; other studies have suggested that this part of the brain is involved in understanding someone else's perspective. The scans also revealed a fall in the activity of the ventromedial prefrontal cortex (vMPFC), an area that helps to compare new information against our own experiences. These two reactions fit with the results of other studies, which suggest that we understand someone else's state of mind by simulating what they are thinking, while suppressing our own perspective so it does not cloud our reasoning. But neither the precuneus nor the vMPFC showed any change in the group who watched the computer. And the

connections between the two areas were weaker in the students who watched the computer compared to those who saw the humans.

The differences weren't for lack of deductive effort; when the students were asked to work out the colour of the problem-solver's hat for themselves, the scans showed equally strong activation in the brain's deductive reasoning centres, regardless of whether the students were watching human or machine. It seems that the technique of placing yourself in someone else's shoes does not apply to artificial intelligence. Because we are aware that robots and computers are controlled by programmes, we do not try to simulate their artificial minds – instead, we judge them by their actions. Indeed, when Ge and Han gave the students the simpler task of just saying what hat colour the problem-solver can see, those watching the computer showed stronger activity in the visual cortex than those watching the humans. That suggests they were paying closer attention to the details of the scene such as where the computer's camera was pointing. Their precuneus, however, remained unexcited.

These results may help to explain why autistic people seem to enjoy interacting with computers and playing with robots. Autistics face social difficulties because they find it hard to put themselves in other people's shoes. Indeed, their vMPFCs fail to tune down in the normal way, suggesting that they cannot block their own experiences from interfering with their deductions about someone else's. But when they interact with robots, they do not have to do that – remember that the activity of the vMPFC did not drop either in the students who watched the problem-solving computers. Ge and Han conclude that humans understand artificial intelligence and other humans using very different mental strategies but it is perhaps a bit premature to make that conclusion based on a simple camera mounted on a computer that never actually interacted with the study's participants. Would the results be different if the robot in question was more human in design, as more and more artificial intelligence is becoming? What would happen in the precuneus and vMPFC of a person playing with a Robo Sapien toy (*above*) or watching WALL-E on the big screen? A question for next time, perhaps.

The rose-tinted cortex

In 1979, a crucified Eric Idle advised movie-goers to always look on the bright side of life. It seems that he needn't have bothered. Psychological experiments have consistently shown that as a species, our minds are awash with a pervasive optimism. We expect our future successes to overpower our past ones. Compared to an imaginary Joe or Jane Bloggs, we deem ourselves more likely to enjoy a long life, more likely to have a successful career and less likely to suffer divorce or ill health. Even the most cynical of minds had a tendency for making similar, overconfident predictions. Now, Tali Sharot and colleagues from New York University have pinpointed a neural circuit in the brain that generates this glass-half-full outlook.[85]

Sharot asked 18 recruits to remember past events or imagine future ones based on on-screen cues (such as "the end of a relationship" or "winning an award"). She then asked them to describe their imaginings along several different lines, like how positive, vivid and emotionally affecting they were, and whether they experienced the event first-hand or observed it from afar. Finally, each person completed a standard questionnaire to score how optimistic they are. Their thoughts bore the clear signs of an optimistic bias. They rated future happy events more positively than past ones and they imagined that windfalls would happen much sooner than negative events would. They also conjured up happy future events from a first-hand viewpoint, while they were more likely to see sad future events from an outsider's perspective.

While the volunteers daydreamed away, Sharot was busy scanning their brains with a technique called functional magnetic resonance imaging (fMRI). She identified two parts of the brain that were more strongly activated when they envisaged positive future events compared to negative ones – the rostral anterior cingulated cortex, or RACC, and the right amygdala. The amygdala is the link between our emotions and our higher brain functions like memory and decision-making. It paints our memories with emotional colours and Sharot thinks her data show that it also allows us to simulate the emotional events of tomorrow.

The amygdala's contributions are moderated by the RACC, a region previously linked to acts of self-reflection, like thinking about preferences, hopes or dreams. Sharot's brain scans revealed that the RACC and amygdala were strongly linked when volunteers imagined happy future events, but not negative ones. And the RACC was more strongly activated in volunteers who scored higher in the optimism questionnaire.

Sharot believes that the RACC helps us to imagine a future event by assessing and summing up the emotions and experiences from out past. But it puts a positive spin on things and tunes down any negative emotional responses from the amygdala. Thanks to the RACC, our past may be writ, but our future is a blank slate where we can happily distance ourselves from negative experiences and move towards positive ones.

Seeing the future through rose-tinted glasses may be a bit naïve, but it is also adaptive. A tendency to expect successful outcomes could provide us with a greater impetus for achieving our goals. While extreme optimism can lead us to harm by underestimating risks, giving too much credence to negative predictions can impair our daily lives.

By identifying the neural circuits involved in optimism, Sharot may also have shed some light on its opposite number – depression. Depression is associated with pessimism and an inability to view the future in detail. It could be that the circuit connecting the RACC to the amygdala is faulty is the brains of depressed people, so that they cannot downplay negative experiences when thinking about the future.

The repressed side of doctors

Many patients would like their doctors to be more sensitive to their needs. That may be a reasonable request but at a neurological level, we should be glad of a certain amount of detachment. Humans are programmed, quite literally, to feel each others' pain. The neural circuit in our brains that registers pain also fires when we see someone else getting hurt; it is why we automatically wince. This empathy makes evolutionary sense – it teaches us to avoid potential dangers that our peers have helpfully pointed out to us. But it can be liability for people like doctors, who see pain on a daily basis and are sometimes forced to inflict it in order to help their patients. Clearly, not all doctors are wincing wrecks, so they must develop some means of keeping this automatic response at bay.

That is exactly what Yawei Chang from Taipei City Hospital and Jean Decety from University of Chicago found. They showed that experienced doctors learn to control the part of their brain that allows them to empathise with a patient's pain, and switch on another area that allows them to control their emotions.[86]

They compared the brains of 14 acupuncturists with at least two years of experience to control group of 14 people with none at all. They scanned the participants' brains while they watched videos of people being pricked by needles in their mouths, hands and feet, or being prodded with harmless cotton swabs. Sure enough, the two groups showed very different patterns of brain activity when they watched the needle videos, but not the cotton swab ones. The controls showed increased brain activity in areas involved in processing pain, including the anterior insula, anterior cingulated cortex, somatosensory cortex and the periaqueductal gray (or PAG). The PAG in particular is thought to act like a control centre for our panic response, mobilising our limbs into action when we sense danger.

In the brains of the more detached experts, these areas failed to light up. Instead, they showed increased activity in parts of the brain involved in higher functions like self-awareness and emotional regulation, including the parahippocampal gyrus, the middle frontal gyrus, the medial prefrontal cortex and the inferior parietal lobule. Even a needle in the mouth did not provoke an empathic pain response in the experts, while these were most likely to activate the somatosensory cortex in the controls. The participants' own answers to a questionnaire backed up the brain scans. When asked to rate the pain and unpleasantness shown in the needle videos, the controls gave an average score of 6.5 out of 10, while the acupuncturists felt that the clips only warranted a three or a four.

Yawei and Decety ruled out the possibility that the doctors were just more emotionally numb than their controls, or more cavalier in the face of pain. In response to psychological questionnaires, the two groups showed no differences in their general levels of empathy or their sensitivity to pain. Instead, the activation pattern in the experts' brains suggests that their higher brain functions block the activation of the pain circuit. The parahippocampal gyrus is partially responsible for retrieving memories – its activation allowed the experts to use their previous years of experience to influence their current reactions. The temporoparietal junction is important in maintaining a distinction between oneself and others, and its activation could allow them to emotionally detach from their patients. With experience, the acupuncturists know that their actions are painful for their patients and have learned to repress their own automatic responses to the sight of pain. That may sound cold, but it is essential for operating a successful practice.

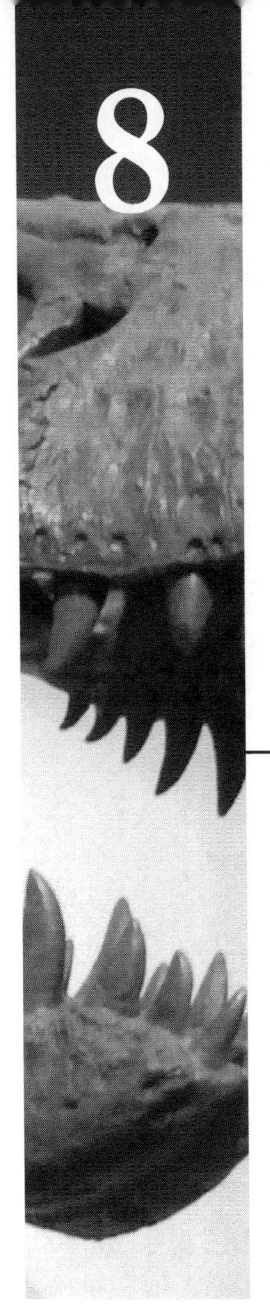

Tales from the grave

Lucky dinosaurs, super-sharks, mammoth-killers and resurrected seeds

The strongest bite in history

Thanks to Hollywood, the jaws of the great white shark may be the most famous in the animal kingdom. But despite its presence in film posters, the great white's toothy mouth received very little experimental attention until Stephen Wroe from the University of New South Wales put it through a digital crash-test, to work out just how powerful its bite is.[87] A medium-sized great white, 2.5m in length and weighing in at 240kg, can bite with a force of 3,000N (0.3 tonnes). But the largest individuals can exert a massive 18,000N (1.8 tonnes) with their jaws, giving them one of the most powerful bites of any living animal. The jaws exert over three times more force than the 5,600N exerted by a large lion, and 20 times more than the 800N a feeble human jawbone can manage.

Impressive as the great white shark is, one of its extinct ancestors was even more so. Megalodon (aka the megatooth shark aka *Carcharadon megalodon*), was a monster that may have grown to 16 metres in length and had a maximum weight of anywhere from 50 to 100 tonnes. And according to Wroe's research, it had the most powerful bite of any animal. A single chomp could exert up to 180,000N (18 tonnes) of force; even the mighty *Tyrannosaurus rex* could only muster 3,000N of force. Being bitten by a Megalodon would be like having three African elephants pressing on top of you with carving knives strapped to their feet. It truly was one of the most powerful predators in history.

If anything, these figures are underestimates. Wroe only modelled what would happen if the sharks raise their lower jaw, whereas an actual bite consists of lots of different movements – for example, a great white's upper jaw pushes forward when it chomps and its head presses downward. Nor do great whites bite in an elegant, genteel way. They frequently ram their prey, driving their jaws towards them at high speed and once bitten, the victim is shaken from side to side. So it is likely that an animal being bitten by these predators would experience forces far larger than the already considerable ones measured in Wroe's study.

Their size is certainly a factor in the strength of their jaws – pound for pound, a lion's bite is actually twice as powerful. Even so, the shark's power is particularly remarkable because it works with flimsier material than a lion or a dinosaur. Shark skeletons are made of cartilage rather than bone, a material that is much more pliant and easily deformed. Even with this disadvantage, the great white and its extinct cousin had no problems in biting hard. They also made up for any structural problems with strategy. Biologists have reported that great white sharks sometimes ambush sea-lions with crippling bites and wait for their victims to bleed out. Megalodon

probably used a similar technique, although it went after much larger prey – whales. Several bones from fossil whales are studded with tooth marks that match the giant gnashers of Megalodon. The wounds tend to be concentrated around the flipper bones and the tail vertebrae, suggesting that the giant shark hunted by first crippling its prey, just as its descendents do in today's oceans.

Wroe describes the great white as an "experimentally intractable" animal, which is a nice way of saying that it is very difficult to run tests on live sharks because of the imminent risk of being eaten. In lieu of live tests, his team scanned the skull of a great white to create a virtual model that they could analyse in safer conditions. They used a technique called 'finite element analysis, which engineers commonly employ to test the mechanical properties of man-made objects; it is used, for example, to see how cars crumple during crash-tests. Wroe's group have taken this technique away from the world of metal and plastics and applied it to the behaviour of bone and cartilage. The models are incredibly realistic, accounting for the different materials that make up the skull, the points that muscles connect with bone, and the ways that the bones themselves fit together. The team have built an impressive portfolio of research around the use of finite element analysis to examine the skulls of some of the planet's most awesome predators.

Their portfolio also includes another of the most famous hunters in prehistory – the sabretooth cat. There is no question that it was a formidable hunter, capable of bringing down large prey like giant bison, horses, and possibly even mammoths. The two massive canines are a striking visual but while they were clearly powerful weapons, scientists have spent over 150 years debating how they were used. Wroe's work shows that *Smilodon*, the most iconic of the sabretooths, had a surprisingly weak bite. Its lengthy canines were precision weapons that were used to deliver a single, final wound to an already subdued victim – the equivalent of an assassin's stiletto rather than a swordsman's blade. Earlier suggestions pictured *Smilodon* using its teeth to hang onto the back of large prey, to slash their abdomens open, or to impale them at the end of a flying pounce. One of the most popular theories said that the cat would have used its teeth to sever arteries and airways with a decisive bite to the throat – a quicker technique than the suffocating neck bites used by modern lions. Working out how strongly *Smilodon* could bite would go a long way towards settling on one of these theories and to do that, palaeontologists have studied the animal's fossilised skull. Even then, opinions have gone either way depending on which bit of the skull they looked at. The muscle attachment points suggest that it had small jaw muscles, but the bite could have been driven from the powerful neck. The lower jaw was smaller than the upper, but strongly built, lending weight to the idea of a powerful bite.

To get some clearer answers, Wroe, together with Colin McHenry and others from the University of Newcastle, Australia analysed *Smilodon*'s skull using finite element analysis.[88] As with the sharks, they used a CT scanner to model the skulls of a *Smilodon* from the famous La Brea Tar Pits in Los Angeles, and a lion from the Taronga Zoo in Sydney. With both predators weighing in fairly equally (the lion was particularly large and the sabre-tooth particularly small), McHenry assumed that they would have tackled similarly sized prey. The team also simulated the jaw and neck muscles of both animals, and put them through a series of simulated bites on the computer. In terms of skull power, the lion outclassed *Smilodon* in almost every way. The 'king of beasts' happily chomped down with a force of over 3,000N (0.3 tonnes), but *Smilodon* only managed a measly 1,000N (0.1 tonnes). Its jaws were remarkably under-powered for a cat of its large size and bulk, biting with the same amount of force as a jaguar about a third of its size. When the powerful neck muscles were factored in, *Smilodon*'s bite force increased to a more respectable 2,000N (0.2 tonnes) – clearly, this was a predator that bit from the neck. But even with its restored reputation, McHenry's analysis also found that *Smilodon* could not have bitten prey on the run. Lions will frequently latch onto running buffalo, and their skulls are built to handle the massive forces that would push and twist against them. Even with 2000N (200kg) of force pushing sideways on a lion's canines, the teeth experience minimal stress. Not so with *Smilodon* – subjected to the same forces, its entire skull experienced tremendous stress and strain. If it tackled large prey that was still on its feet, it would have run a strong risk of snapping its teeth or breaking its skull in multiple places.

The sabre-toothed cat's relatively weak bite rules out a lot of the theories for its killing style. It could not possibly have tackled running prey and slashing at the belly would have left it vulnerable to snapping teeth if the victim tried to get back up. *Smilodon*'s only real option was to use its teeth to deliver a killing bite when there was no chance of the prey actually moving. It was a one-use weapon, reserved for prey that had been brought down and pinned, preferably at the head, by the cat's enormous bulk. Fortunately, the rest of the animal was superbly adapted for this. *Smilodon* had a physique that was more bear than cat and it had over-sized 'dew claws' on its 'thumbs'. All of these traits would have given it enough power and inertia to bring down large animals in a way that modern lions simply cannot do. Lions often kill with a lengthy suffocating bite that can last for up to 13 minutes. In *Smilodon*'s case, the fate of the downed animal would have been sealed within seconds, especially if the sabre teeth severed the carotid artery.

McHenry speculates that *Smilodon*'s technique could have evolved because of intense competition from other large predators, such as the fearsome dire wolves, the short-faced bear and the immense American lion. With competitors like these on the prowl and ready to fight over a carcass, *Smilodon* had little time for a long drawn-out kill. And while its bulk and huge teeth would have made it well adapted for hunting large prey, McHenry thinks that that *Smilodon,* unlike modern cats, would have been woefully inadequate at killing smaller and more agile animals. When the large herbivores of the last Ice Age all died out, *Smilodon* would have been unable to compete with more versatile killers.

Outgrowing your enemies

A good defence was a vital part of life in the Cretaceous. Plant-eating dinosaurs needed effective ways of warding off the crushing jaws of *Tyrannosaurus* and its kin. Some species like *Triceratops* and *Ankylosaurus* had fairly obvious protective equipment, including horns, frills and armoured plates. But others lacked defensive armaments and had to fend off predators through subtler means. Take *Hypacrosaurus.* It was one of the duck-billed dinosaurs known as hadrosaurs, and like most other members of the group, its soft body lacked any obvious protection. Its main advantage was size; a fully-grown adult was an immense animal that almost rivalled *T.rex* in height. Its name even means "near the highest lizard". And if a large size is your only defence, it is a good idea to grow quickly.

That is exactly what *Hypacrosaurus* did. Lisa Noelle Cooper from Kent State University has shown that the dinosaur reached its towering size in record time and grew much faster than the predators that hunted it.[89] Other dinosaurs sought refuge behind shields and armour, but this otherwise

defenceless species hid in plain sight, behind a large bulk attained at an extraordinary rate.

Together with Andrew Lee, Mark Taper and legendary palaeontologist Jack Horner, Cooper looked at the telltale rings in the leg bones of one *Hypacrosaurus* specimen. Like the rings of trees, these bands represent a year of growth. By analysing them, Cooper could reconstruct the early years of this long-dead individual. She calculated that the animal was about 13 years old when it died but the increasingly narrow spacing of its growth rings showed that it had clearly finished growing. The rings also revealed that the hadrosaur went through a massive early growth spurt that catapulted it to full size in just 10-12 years.

Hypacrosaurus shared its homelands with four predators, including *Troodon*, a small but intelligent hunter, and no less than three tyrannosaurs – *Albertosaurus*, *Daspletosaurus* and *Gorgosaurus*. To see if its fast growth was a defence against these hunters, Cooper compared *Hypacrosaurus*'s speedy spurt with those of *Albertosaurus* and *Tyrannosaurus* (standing in for *Daspletosaurus*). She found that unlike *Hypacrosaurus*, the giant predators were late bloomers; *Albertosaurus* took 23 years to reach its full stature, while *Tyrannosaurus* took 36 years to crawl to adulthood. That means that *Hypacrosaurus* grew 2-4 times faster than the predators that hunted it. Even though *Albertosaurus* would eventually outsize its prey, it was only half-sized by the time its quarry was fully grown.

It is difficult to generalise to an entire species from a single specimen, but these results fit with earlier estimates using other duck-billed dinosaurs.

For example, a closely related species called *Maiasaura* is thought to have put on over a tonne of weight every year during it first few years of life. This pattern of growth is what you would expect of hunted species, whose young are likely to die early at the jaws of predators. That was certainly the case for *Hypacrosaurus* and other hadrosaurs – the bones of juveniles have been found *inside* the skeleton of a *Daspletosaurus*, and one *Hypacrosaurus* leg bone actually has a tooth of a predatory dinosaur embedded inside it! In response to these dangers, species tend to grow quickly and start reproducing early. Indeed, the growth of Cooper's *Hypacrosaurus* sped up dramatically in the first 2-3 years after hatching and subsequently slowed down. Recent studies suggest that this turning point marks the age at which dinosaurs become sexually mature, so this creature was probably fertile just a few years after emerging from its egg.

Resurrection of the *Phoenix* seed

On an isolated rock plateau in southern Israel, surrounded by steep cliff-faces, stands the imposing fortress of Masada. The site was built by King Herod as a pleasure palace and became famous for the alleged mass suicide of its Jewish inhabitants, who chose death over capture by Roman invaders. But while the fortress's occupants have long died, one former resident is still around – an ancient seed that survived beneath the rubble for 2,000 years and has today germinated into a tree. The seed comes from a date palm and appropriately enough, its carries the genus name of *Phoenix* and the nickname of "Methuselah".

The seed was one of three that were excavated from Masada in the 1960s. It was stored at room temperature for four decades before being finally planted in 2005 by a team led by Sarah Sallon at the Hadassah Medical Organisation in Jerusalem. The event took place on January 19, the date of the Jewish New Year of Trees, when new trees are traditionally planted. Eight weeks later, it had started to sprout. Under the careful attention of plant specialist Elaine Solowey, the tree developed in the normal way. At first, its early leaves developed white patches where they lacked chlorophyll, probably because of a nutrient deficiency. But Methuselah managed to shrug off these early problems, and by early 2007, it was four feet tall.

Sallon confirmed the seed's extreme age by carbon-dating small fragments of its original shell that were still clinging to its roots.[90] The technique assesses the levels of a radioactive form of carbon that accumulates in living tissues but starts to fall in a predictable way once an organism dies. Based on these measurements, Sallon found that the seed was two millennia old, give or take about 50 years. The resurrected date palm smashes the previous record for oldest seed ever germinated, held by a 1,300-year-old lotus seed. Sallon thinks that its longevity is the result of the high summer temperatures and low levels of rainfall at Masada. These harsh conditions dried the seed and curtailed the production of free radicals, molecules that are a major force in seed ageing.

The date palm (*Phoenix dactylifera*) was first domesticated about 5,000 years ago and at the time that the Methuselah seed was first produced, the Judean Dead Sea region was famous for its extensive trade in dates. Since then, the historic date populations have been lost. Today, only a few, low-quality palms are propagated from local seeds. Israel imports most of its date palms. Methuselah, however, represents a second coming for this extinct population. It shares only half of its genetic material with modern dates from Morocco, Egypt and Iraq, and the remainder could allow geneticists to peer back in time at their genes. If Methuselah turns out to be female (and Sallon will only know that for certain in about two years), it may even help to restore the date palms of old.

Lucky dinosaurs

Some 230 million years ago, giant reptiles walked the Earth. Some were large and fearsome predators; others were nimble and fleet-footed runners; and yet others were heavily armoured with bony plates running down their backs. Their bodies had evolved into an extraordinary range of shapes and sizes and they had done so at a breakneck pace. They were truly some of the most impressive animals of their time. They were the crurotarsans.

Wait... the what now? Chances are you have never heard of the crurotarsans and you were expecting that other, more famous group of giant reptiles – the dinosaurs. There is certainly no doubt that the dinosaurs were an evolutionary success story, diversifying from a standard body plan – a small, two-legged meat-eater – into a dazzling selection of forms. Today, living dinosaurs – the birds – still rule our skies and back in their heyday, they were the dominant back-boned animals on land for millions of years. But what was the secret to their rise to power? Many palaeontologists believe that they simply outcompeted other animal groups that were around at the same time like the bizarre, buck-toothed rhynchosaurs or mammal-

A selection of crurotarsans.
Top row, from left to right: Effigia, Rutiodon, Desmatosuchus.
Bottom row, from left to right: Postosuchus, Lotosaurus, Shuvosaurus

like reptiles like the cynodonts. Perhaps their upright postures made them faster or more agile; perhaps they were actually warm-blooded and able to cope with a wider range of climates. Either way, the fact that they and not the other reptile groups ascended to dominance is often taken as a sign of their superiority.

Stephen Brusatte from Columbia University challenges that view.[91] According to his take on Triassic life, the dinosaurs fared no better than their competitors, the crurotarsans, and were actually less successful for about 30 million years. They eventually supplanted these other groups because of luck rather than because they possessed any special advantage. During the Triassic period, the crurotarsans (which eventually gave rise to today's crocodiles and alligators) were at their most diverse. They ranged from top predators like *Postosuchus* to the armoured giants like *Desmatosuchus* to swift, two-legged runners like *Effigia* and *Shuvosaurus*. Many of them were incredibly similar to the dinosaurs we know and love (see top image) and some were even mistaken for dinosaurs when they were first discovered. Their strikingly similar bodies suggest that members of the two groups shared similar lifestyles and probably competed for the same resources. Did the dinosaurs simply win the battle? Brusatte noted that answers to this question often invoke nebulous concepts of "superiority" and he wanted to look at it from a more objective angle, comparing the two groups along measurable lines. To that end, he worked with three other palaeontologists to construct a family tree of 64 dinosaurs and crurotarsans, based on 437 features on each of their skeletons.

The team calculated how fast each group was evolving based on their bones. If dinosaurs really outcompeted the crurotarsans, you would expect to see the former group evolving at increasing rates during the Triassic, while the latter group's rate of evolution slowed. But that is not what happened. Instead, Brusatte found that during the Triassic as a whole, the crurotarsans were keeping pace with the expansion of the dinosaur lineage. It is possible that during the mid-Triassic, the dinosaurs were evolving at a slightly more rapid rate, but it is hard to be sure based on such few samples. Brusatte also looked at the range of body plans developed by each group, also known as their "disparity". Animal groups with high disparity and a wide range of body shapes tend to have a more varied set of lifestyles, habitats and diets. And surprisingly, the crurotarsans had about twice as much disparity as the dinosaurs did at the time.

So for the 30 million years when the dinosaurs and the crurotarsans shared the planet, both were evolving at equal rates and it was the crurotarsans who were experimenting with twice as many body shapes. To Brusatte, that is a blow to the long-standing view that the "superior" dinosaurs were in some way "preordained for success". Instead, he suggests that the dinosaurs' triumph hinged on a combination of "good luck" and perseverance. Both they and the crurotarsans survived an extinction event 228 million years ago, which wiped out many other reptile groups like the rhynchosaurs. At the end of the Triassic period, some 28 million years later, the dinosaurs weathered another (much bigger) extinction event that finally killed the majority of crurotarsans off for good.

It is not clear why the dinosaurs persevered and the crurotarsans did not. Perhaps they had some unique adaptation that the crurotarsans lacked, which ensured their survival. But Brusatte says that this explanation is "difficult to entertain" because the crurotarsans were more abundant at the time and had vastly more varied bodies. He also says that the death of certain groups during mass extinctions is more likely to be due to random factors rather than any specific aspect of their lifestyles. Whatever the answer, the sudden *volte face* of the crurotarsans, from ruling reptiles to evolutionary footnotes, gave the dinosaurs their chance. They were the reptilian equivalent of videotapes, rising to dominance in the wake of a superior Betamax technology. In the brave, new world of the Jurassic, they could exploit the niches vacated by their fallen competitors.

The rise of the dinosaurs is often spoken of as a single event but it was more likely to have been a two-stage process. The predecessors of the gigantic, long-necked sauropods expanded into new species after the late Triassic extinction, while the big meat-eaters and the armoured plant-eaters only came to the fore as a second extinction heralded the start of the Jurassic. Brusatte refers to the dinosaurs as the "beneficiaries of two mass extinction events", which is ironic given what happened next. About a 130 million years later, the dinosaurs' luck proved to be finite. They survived through two extinctions, but as the saying goes, third time's the charm.

What killed the mammoths?

The image of cavemen hunting mammoths is a popular one in Hollywood, but did our ancestors really exterminate the woolly mammoth? Well, sort of. It turns out that humans only delivered a killing blow to a species that had already been driven to the brink of extinction by changing climates. Corralled into a tiny range by habitat loss, the diminished mammoth population became vulnerable to the spears of hunters. We just kicked them while they were down.

The woolly mammoth first walked the earth about 300,000 years ago during the Pleistocene period. They were well adapted to survive in the dry and cold habitat known as the 'steppe-tundra'. Despite the sparse plant life there, the woolly mammoths were very successful, spreading out in a belt across the Northern hemisphere. Their fortunes began to change as the Pleistocene gave way to the Holocene. The climate around them started to become warmer and wetter and the shrinking steppe-tundras greatly reduced the mammoth's habitats. The species made its last stand on the small Wrangel Island in Siberia before finally succumbing to extinction. But climate change is not the whole story. About 40,000 years ago, those relentless predators – human beings – started encroaching into the woolly mammoth's range in northern Eurasia. Which of these two threats, climate change or human hunters, sealed the mammoth's fate?

Climate change has been blamed in the past, but some scientists have noted that the species had survived through other warm spells, including a period about 126,000 years ago that shrunk its habitat to a larger extent than the Holocene warming. The arrival of humans also fits the general timing, but some sources claim that there is little direct evidence of our ancestors actually hunting mammoths. David Nogues-Bravo from the National Museum of Natural Sciences in Madrid thinks that the two factors worked together, with climate changes weakening the population and humans finishing it off.[92] Taking factors like temperature and rainfall into account, he ran simulations of the mammoth's prehistoric climate at five time points over a span of 120,000 years. These models were bolstered by data from the carbon-dated fossil remains of woolly mammoths that lived throughout Eurasia during this time period. The simulations showed that the area of land containing suitable habitats for the mammoths was incredibly sparse about 126,000 years ago. At that point, the giants were most likely confined to the very north of Eurasia but as the climate cooled, they spread outwards and their geographical range increased by 26 times to encompass much of the northern continent. The turning point happened about 42,000 years ago when their habitat started to shrink catastrophically. Over the next 36,000 years, the species had been relegated to an area just 11% of its former size but even so, their range was still larger than it was at the previous low.

Nogues-Bravo then modelled how the mammoths that were still around 6,000 years ago would have reacted to being hunted by humans where their ranges overlapped. He found that even with an optimistic estimate of the population's size and strength, each human would have had to kill just one mammoth every three years to drive the species to extinction. With a pessimistic appraisal of mammoth numbers, the death of just one mammoth every 200 years at the hands of each human would have been enough to seal their fate. If humans had never arrived in northern Eurasia, the mammoths might still have made it. Nogues-Bravo suggests that they would probably have survived in small pockets of suitable habitat and made use of fringing and less welcoming areas, as they probably did during the warm spell 126,000 years ago. But they never got the chance. Our ancestors delivered the coup de grace to a failing population that was too small to withstand the hunting pressures they were subjected to.

References

1. Inoue and Matsuzawa. (2007) Working memory of numerals in chimpanzees. Curr Biol 17(23): R1004-5.
2. Begall, et al. (2008) Magnetic alignment in grazing and resting cattle and deer. Proc Natl Acad Sci U S A
3. Gal and Libersat. (2008) A parasitoid wasp manipulates the drive for walking of its cockroach prey. Curr Biol 18(12): 877-82.
4. Grosman, et al. (2008) Parasitoid increases survival of its pupae by inducing hosts to fight predators. PLoS ONE 3(6): e2276.
5. Vaughn and Strathmann. (2008) Predators induce cloning in echinoderm larvae. Science 319(5869): 1503.
6. Blackiston, Silva Casey, and Weiss. (2008) Retention of memory through metamorphosis: can a moth remember what it learned as a caterpillar? PLoS ONE 3(3): e1736.
7. de Waal, Leimgruber, and Greenberg. (2008) Giving is self-rewarding for monkeys. Proc Natl Acad Sci U S A 105(36): 13685-9.
8. Blackburn, Hanken, and Jenkins. (2008) Concealed weapons: erectile claws in African frogs. Biol Lett 4(4): 355-7.
9. King, Douglas-Hamilton, and Vollrath. (2007) African elephants run from the sound of disturbed bees. Curr Biol 17(19): R832-3.
10. Bazazi, et al. (2008) Collective motion and cannibalism in locust migratory bands. Curr Biol 18(10): 735-9.
11. Jonsson, et al. (2008) Tardigrades survive exposure to space in low Earth orbit. Curr Biol 18(17): R729-31.
12. Gladyshev and Meselson. (2008) Extreme resistance of bdelloid rotifers to ionizing radiation. Proc Natl Acad Sci U S A 105(13): 5139-44.
13. Miserez, et al. (2008) The transition from stiff to compliant materials in squid beaks. Science 319(5871): 1816-9.
14. Kastberger, Schmelzer, and Kranner. (2008) Social waves in giant honeybees repel hornets. PLoS ONE 3(9): e3141.
15. Joly-Mascheroni, Senju, and Shepherd. (2008) Dogs catch human yawns. Biol Lett 4(5): 446-8.
16. Chiou, et al. (2008) Circular polarization vision in a stomatopod crustacean. Curr Biol 18(6): 429-34.
17. Bates, et al. (2007) Elephants classify human ethnic groups by odor and garment color. Curr Biol 17(22): 1938-42.

18. Karsten, et al. (2008) A unique life history among tetrapods: an annual chameleon living mostly as an egg. Proc Natl Acad Sci U S A 105(26): 8980-4.
19. Wiens, et al. (2008) Chronic intake of fermented floral nectar by wild treeshrews. Proc Natl Acad Sci U S A 105(30): 10426-31.
20. Friedman. (2008) The evolutionary origin of flatfish asymmetry. Nature 454(7201): 209-12.
21. Kozmik, et al. (2008) Assembly of the cnidarian camera-type eye from vertebrate-like components. Proc Natl Acad Sci U S A 105(26): 8989-93.
22. Thewissen, et al. (2007) Whales originated from aquatic artiodactyls in the Eocene epoch of India. Nature 450(7173): 1190-4.
23. Pouchkina-Stantcheva, et al. (2007) Functional divergence of former alleles in an ancient asexual invertebrate. Science 318(5848): 268-71.
24. Gladyshev, Meselson, and Arkhipova. (2008) Massive horizontal gene transfer in bdelloid rotifers. Science 320(5880): 1210-3.
25. Blount, Borland, and Lenski. (2008) Historical contingency and the evolution of a key innovation in an experimental population of Escherichia coli. Proc Natl Acad Sci U S A 105(23): 7899-906.
26. Simmons, et al. (2008) Primitive Early Eocene bat from Wyoming and the evolution of flight and echolocation. Nature 451(7180): 818-21.
27. Paegel and Joyce. (2008) Darwinian evolution on a chip. PLoS Biol 6(4): e85.
28. Hanifin and Brodie. (2008) Phenotypic mismatches reveal escape from arms-race coevolution. PLoS Biol 6(3): e60.
29. Atkinson, et al. (2008) Languages evolve in punctuational bursts. Science 319(5863): 588.
30. Lieberman, et al. (2007) Quantifying the evolutionary dynamics of language. Nature 449(7163): 713-6.
31. Kirby, Cornish, and Smith. (2008) Cumulative cultural evolution in the laboratory: an experimental approach to the origins of structure in human language. Proc Natl Acad Sci U S A 105(31): 10681-6.
32. Tewksbury, et al. (2008) Evolutionary ecology of pungency in wild chilies. Proc Natl Acad Sci U S A 105(33): 11808-11.
33. Munson, et al. (2008) Climate extremes promote fatal co-infections during canine distemper epidemics in African lions. PLoS ONE 3(6): e2545.
34. Kuris, et al. (2008) Ecosystem energetic implications of parasite and free-living biomass in three estuaries. Nature 454(7203): 515-8.

35. Kurz, et al. (2008) Mountain pine beetle and forest carbon feedback to climate change. Nature 452(7190): 987-90.
36. Carpenter, et al. (2008) One-third of reef-building corals face elevated extinction risk from climate change and local impacts. Science 321(5888): 560-3.
37. Russ, et al. (2008) Rapid increase in fish numbers follows creation of world's largest marine reserve network. Curr Biol 18(12): R514-5.
38. Sweatman. (2008) No-take reserves protect coral reefs from predatory starfish. Curr Biol 18(14): R598-9.
39. Fomina, et al. (2008) Role of fungi in the biogeochemical fate of depleted uranium. Curr Biol 18(9): R375-7.
40. Palmer, et al. (2008) Breakdown of an ant-plant mutualism follows the loss of large herbivores from an African savanna. Science 319(5860): 192-5.
41. Wu, et al. (2008) Suppression of cotton bollworm in multiple crops in China in areas with Bt toxin-containing cotton. Science 321(5896): 1676-8.
42. Bliege Bird, et al. (2008) The "fire stick farming" hypothesis: Australian Aboriginal foraging strategies, biodiversity, and anthropogenic fire mosaics. Proc Natl Acad Sci U S A 105(39): 14796-801.
43. Christner, et al. (2008) Ubiquity of biological ice nucleators in snowfall. Science 319(5867): 1214.
44. La Scola, et al. (2008) The virophage as a unique parasite of the giant mimivirus. Nature 455(7209): 100-4.
45. Rambaut, et al. (2008) The genomic and epidemiological dynamics of human influenza A virus. Nature 453(7195): 615-9.
46. Russell, et al. (2008) The global circulation of seasonal influenza A (H3N2) viruses. Science 320(5874): 340-6.
47. Mendell, et al. (2008) Extreme polyploidy in a large bacterium. Proc Natl Acad Sci U S A 105(18): 6730-4.
48. Dantas, et al. (2008) Bacteria subsisting on antibiotics. Science 320(5872): 100-3.
49. Halfmann, et al. (2008) Generation of biologically contained Ebola viruses. Proc Natl Acad Sci U S A 105(4): 1129-33.
50. Chivian, et al. (2008) Environmental genomics reveals a single-species ecosystem deep within Earth. Science 322(5899): 275-8.
51. Yu, et al. (2008) Neutralizing antibodies derived from the B cells of 1918 influenza pandemic survivors. Nature 455(7212): 532-6.
52. Prabhakar, et al. (2008) Human-specific gain of function in a developmental enhancer. Science 321(5894): 1346-50.

53. Spalding, et al. (2008) Dynamics of fat cell turnover in humans. Nature 453(7196): 783-7.
54. Narkar, et al. (2008) AMPK and PPARdelta agonists are exercise mimetics. Cell 134(3): 405-15.
55. Takahashi, et al. (2007) Induction of pluripotent stem cells from adult human fibroblasts by defined factors. Cell 131(5): 861-72.
56. Yu, et al. (2007) Induced pluripotent stem cell lines derived from human somatic cells. Science 318(5858): 1917-20.
57. Dimos, et al. (2008) Induced pluripotent stem cells generated from patients with ALS can be differentiated into motor neurons. Science 321(5893): 1218-21.
58. Waterland, et al. (2008) Methyl donor supplementation prevents transgenerational amplification of obesity. Int J Obes (Lond)
59. Walum, et al. (2008) Genetic variation in the vasopressin receptor 1a gene (AVPR1A) associates with pair-bonding behavior in humans. Proc Natl Acad Sci U S A
60. Audero, et al. (2008) Sporadic autonomic dysregulation and death associated with excessive serotonin autoinhibition. Science 321(5885): 130-3.
61. Cant and Johnstone. (2008) Reproductive conflict and the separation of reproductive generations in humans. Proc Natl Acad Sci U S A 105(14): 5332-6.
62. Helgason, et al. (2008) An association between the kinship and fertility of human couples. Science 319(5864): 813-6.
63. Hagemann, Strauss, and Leibing. (2008) When the referee sees red... Psychol Sci 19(8): 769-771.
64. Galdi, Arcuri, and Gawronski. (2008) Automatic mental associations predict future choices of undecided decision-makers. Science 321(5892): 1100-2.
65. Dunn, Aknin, and Norton. (2008) Spending money on others promotes happiness. Science 319(5870): 1687-8.
66. Karpicke and Roediger. (2008) The critical importance of retrieval for learning. Science 319(5865): 966-8.
67. Herrmann, Thoni, and Gachter. (2008) Antisocial punishment across societies. Science 319(5868): 1362-7.
68. Dreber, et al. (2008) Winners don't punish. Nature 452(7185): 348-51.
69. Smith, et al. (2008) Lacking power impairs executive functions. Psychol Sci 19(5): 441-7.
70. Hassin, et al. (2007) Subliminal exposure to national flags affects political thought and behavior. Proc Natl Acad Sci U S A 104(50): 19757-61.

71. Stetson, Fiesta, and Eagleman. (2007) Does time really slow down during a frightening event? PLoS ONE 2(12): e1295.
72. Kaminski, Sloutsky, and Heckler. (2008) Learning theory. The advantage of abstract examples in learning math. Science 320(5875): 454-5.
73. Zhong and Leonardelli. (2008) Cold and lonely: does social exclusion literally feel cold? Psychol Sci 19(9): 838-42.
74. Williams and Bargh. (2008) Experiencing physical warmth promotes interpersonal warmth. Science 322(5901): 606-7.
75. Limb and Braun. (2008) Neural substrates of spontaneous musical performance: an FMRI study of jazz improvisation. PLoS ONE 3(2): e1679.
76. Jaeggi, et al. (2008) Improving fluid intelligence with training on working memory. Proc Natl Acad Sci U S A 105(19): 6829-33.
77. Di Dio, Macaluso, and Rizzolatti. (2007) The golden beauty: brain response to classical and renaissance sculptures. PLoS ONE 2(11): e1201.
78. Velliste, et al. (2008) Cortical control of a prosthetic arm for self-feeding. Nature 453(7198): 1098-101.
79. Cattaneo, et al. (2007) Impairment of actions chains in autism and its possible role in intention understanding. Proc Natl Acad Sci U S A 104(45): 17825-30.
80. Volkow, et al. (2008) Methylphenidate decreased the amount of glucose needed by the brain to perform a cognitive task. PLoS ONE 3(4): e2017.
81. Soon, et al. (2008) Unconscious determinants of free decisions in the human brain. Nat Neurosci 11(5): 543-5.
82. Shaw, et al. (2007) Attention-deficit/hyperactivity disorder is characterized by a delay in cortical maturation. Proc Natl Acad Sci U S A 104(49): 19649-54.
83. Kay, et al. (2008) Identifying natural images from human brain activity. Nature 452(7185): 352-5.
84. Ge and Han. (2008) Distinct neurocognitive strategies for comprehensions of human and artificial intelligence. PLoS ONE 3(7): e2797.
85. Sharot, et al. (2007) Neural mechanisms mediating optimism bias. Nature 450(7166): 102-5.
86. Cheng, et al. (2007) Expertise modulates the perception of pain in others. Curr Biol 17(19): 1708-13.
87. Wroe, et al. (2008) Three-dimensional computer analysis of white shark jaw mechanics: how hard can a great white bite? J Zool Epub 12 August 2008(

88. McHenry, et al. (2007) Supermodeled sabercat, predatory behavior in Smilodon fatalis revealed by high-resolution 3D computer simulation. Proc Natl Acad Sci U S A 104(41): 16010-5.
89. Cooper, et al. (2008) Relative growth rates of predator and prey dinosaurs reflect effects of predation. Proc Biol Sci 275(1651): 2609-15.
90. Sallon, et al. (2008) Germination, genetics, and growth of an ancient date seed. Science 320(5882): 1464.
91. Brusatte, et al. (2008) Superiority, competition, and opportunism in the evolutionary radiation of dinosaurs. Science 321(5895): 1485-8.
92. Nogues-Bravo, et al. (2008) Climate change, humans, and the extinction of the woolly mammoth. PLoS Biol 6(4): e79.

Image credits

All images sourced from Wikipedia Commons.

Cover photos (from left to right): coral reef by James Walter; The Thinker by CJ; *Tyrannosaurus rex* skull by David Monniaux, peppered moth (*Biston betularia*) by Olaf Leillinger; human embryo by Ed Uthman; chameleon by Ridard; neurons by BrainMaps.org, *Salmonella typhimurium* by Rocky Mountain Laboratories, NIAID, NIH.

Page 4	Cattle (*Bos Taurus*) by Scott Bauer
Page 7	Emerald cockroach wasp (*Ampulex compressa*) by Axel Boldt
Page 10	Sand dollar by Gerhard H
Page 12	Tobacco hornworm (*Manduca sexta*) by Tozeuma
Page 14	Brown capuchins (*Cebus apella*) by Frans de Waal
Page 17	Hairy frog (*Trichobatrachus robustus*) by Gustav Ocarra
Page 22	Tardigrades (*Hypsibus dujardini*) by Willow Gabriel and Bob Goldstein
Page 28	Giant bee (*Apis dorsata*) nest by Sean Hoyland
Page 32	Peacock mantis shrimp (*Odontodactylus scyllarus*) by Jens Petersen
Page 35	Elephants (*Loxodonta africana*) by John Storr
Page 41	Common dab (*Limanda limanda*) by Hans Hillewaert
Page 45	*Indohyus* by Arthur Weasley
Page 47	Bdelloid rotifers by Diego Fontaneto
Page 51	Richard Lenski's long-term evolution experiment by Brian Baer and Neerja Hajela
Page 57	Coast garter snake (*Thamnophis elegans*) by Steve Jurvetson
Page 69	Chillies by Mark Richards
Page 71	Lions (*Felis leo*) by The Lily Breasted Roller
Page 75	Mountain pine beetle (*Dendroctonus ponderosae*) by US Department of Agriculture
Page 78	Staghorn coral (*Acropora cervicornis;* left) by Alessandro Dona; elkhorn coral (*Acropora palmata*) by National Oceanic and Atmospheric Administration
Page 83	Acacia tree (*Acacia collinsi*) with ant by Kurt Stueber
Page 91	Snowflake under an electron microscope by US Department of Agriculture
Page 94	Influenza virus by Cynthia Goldsmith
Page 101	Ebola virus by Frederick Murphy

Index

Acknowledgements

This book is the result of a year's worth of effort, much of it done in the wee hours of the morning. My thanks go out to the various people who have provided me with the motivation to keep Not Exactly Rocket Science alive, including the various bloggers who have linked to the site and helped to spread the word, the good people at ScienceBlogs for giving the blog a new home, and especially to all the readers who have taken the time to read the posts and comment on them. The slew of kind comments, curious questions and chastising critiques have helped me to become a better writer.

At the risk of sounding saccharine, I also want to extend my thanks to the scientists whose work is detailed within these pages. Reading and writing about science gives me a very unique joy and these people have provided the raw material from which the blog and this book have been crafted. Some of it is exciting, some revelatory and some just fun – but all of it is worth knowing about.

Above all else, my beautiful wife Alice has ceaselessly supported my efforts to break into science communication since the very beginning. None of this would be possible without her unremitting support and it is to her that this book is rightly dedicated.